D1553423

PERGAMON INTERNATIONAL LIBRARY
of Science, Technology, Engineering and Social Studies
The 1000-volume original paperback library in aid of education, industrial training and the enjoyment of leisure
Publisher: Robert Maxwell, M.C.

EARTH AND COSMOS

Some other Pergamon titles of interest

UNVEILING THE UNIVERSE
Edited by D. Abir

THE PLANET EARTH
2nd Edition
Edited by D. R. Bates

A STRATEGY FOR THE OZONE LAYER
Edited by A. K. Biswas

INTERNATIONAL GEOGRAPHY – 76
Volume 2 — Climatology, Hydrology, Glaciology
Edited by P. Gerasimov

STELLAR EVOLUTION
2nd Edition
A. J. Meadows

EARTH AND COSMOS

a book relating the Environment of Man on Earth to the Environment of Earth in the Cosmos

BY

ROBERT S. KANDEL

Service d'Aéronomie du Centre National de la Recherche Scientifique, Verrières-le-Buisson, France

PERGAMON PRESS

OXFORD · NEW YORK · TORONTO · SYDNEY · PARIS · FRANKFURT

U.K.	Pergamon Press Ltd., Headington Hill Hall, Oxford OX3 0BW, England
U.S.A.	Pergamon Press Inc., Maxwell House, Fairview Park, Elmsford, New York 10523, U.S.A.
CANADA	Pergamon of Canada Ltd., Suite 104, 150 Consumers Road, Willowdale, Ontario M29 1P9, Canada
AUSTRALIA	Pergamon Press (Aust.) Pty. Ltd. P.O. Box 544, Potts Point, N.S.W. 2011, Australia
FRANCE	Pergamon Press SARL, 24 rue des Ecoles, 75240 Paris, Cedex 05, France
FEDERAL REPUBLIC OF GERMANY	Pergamon Press GmbH, 6242 Kronberg-Taunus, Pferdstrasse 1, Federal Republic of Germany

First edition 1980

British Library Cataloguing in Publication Data

Kandel, R. S.
Earth and Cosmos. – (Pergamon international library).
1. Earth
I. Title
550 QE501 79-40949

ISBN 0-08-025016-5 (Hardcover)
ISBN 0-08-023086-5 (Flexicover)

Printed and bound in Great Britain by
William Clowes (Beccles) Limited
Beccles and London

Contents

Preface

"A universal solidarity links each to all things, and in particular, the Earth to the entire Universe": so wrote Mach, as Robert Kandel reminds us in his third chapter. This principle was as a wellspring of the General Theory of Relativity. For Einstein, Mach's Principle was in essence identical to the principle of equivalence of inertial and gravitational mass, and indeed he went so far as to postulate a specific relation between the constant of universal gravitation, and the cosmic mass distribution (an idea worth pursuing today) — the one being thus a measure of the other.

However, in certain cases, this profound dependence of the local on the universal is neglected by physicists. Need we apprehend the universe, in order to understand properly the forces and interactions between the elementary particles? Einstein thought so, and at his death still was seeking a unified theory of nature. But without even an intuition of the direction of the royal road to unity, we can progress by shortcuts, using empirically determined constants in microphysics, as in macrophysics we use the gravitational constant, which is but a reflection of the distribution of matter in the Universe.

Terrestrial phenomena depend typically on a very large number of physical factors. Lacking even a schematic theory of the totality of the forces acting on the Earth in its solar and galactic environment, one can of course reduce the influence of this environment to a few basic well known parameters, and thenceforth consider the Earth to be a practically closed system. Terrestrial physics then becomes a straightforward affair of calculation, subject to certain boundary and initial conditions. An excellent example of this is to be found in meteorology, even today. In attempting to predict the weather, Man seeks to know how the temperature, cloudiness, winds, rain, . . . will change at one point or another. These are quantities that can be measured day after day, in any particular place. Although the physics behind their evolution may be very complicated, "reasonable", purely local extrapolation of locally observed conditions is always feasible; and this is how one used to go about predicting the weather. What was reasonable not so many years ago, no longer is so today: we now speak of the arrival of warm air masses, the descent of the polar front, the formation of cyclones. We see these phenomena from well-equipped ground stations, and from weather ships at sea; better still, and most completely, we observe them from satellites. And so, reasonable extrapolation can no longer be "local"; in recognizing the dependence of the weather here, on the winds there, we have reached a new level of understanding. Thus in our modern study of the motions of air masses, in scientific dynamic meteorology, we must consider the rotation of the Earth, and even its motion around the Sun.

The Sun heats the Earth, but its heating is variable. . . . Variable first of all because of the complexities of the Earth's orbit, which modulate the distribution of sunlight over its surface. . . . Day and night, winter and summer, and the more or less favored climatic zones, as a function of latitude, of course come to mind. But there also are more complex effects: the precession of the equinoxes, and the perihelion precession, both more or less directly the result of perturbations of the Earth's motions by the Moon and the planets respectively. The time scales of these phenomena (26 000 and 100 000 years) have led many researchers to link them with certain major variations of climate — the ice ages. Finally, the Sun is intrinsically variable, as an active magnetic star. The sunspots — powerful magnetic poles; the flares in active regions — ejecting streams of electrons and protons into space, some towards the Earth; the contortions of the prominences, in the coronal magnetic field; the sometimes steady, sometimes gusty, solar wind: — how could all these *not* affect the minuscule Earth, immersed in the solar sphere of influence? How could these *not* perturb the Earth's outer layers, magnetosphere obviously, ionosphere of course, but also — why not? — the atmosphere: atmosphere — and therefore the weather, when we follow its changes day by day, or even hour by hour? Atmosphere — and therefore the climate, over longer time intervals? The mechanisms of such interactions are, it must be admitted, still obscure; and so it is not yet feasible to include this new component in a further refinement of meteorology. We have already discarded purely local extrapolation, made obsolete by our understanding of the motions of air masses; our dynamic meteorology of today constitutes a more complete extrapolation, assimilating as it does the global and not merely local situation. Still, it assumes a "constant Sun"; will not tomorrow's still more reasonable extrapolation take into account as well what happens outside the Earth, and in particular the activity of the Sun?

The Earth is subject to the laws of astronomy and physics, and for ages Man has been subject to the rigors of the atmosphere, his only recourse before the invading march of ice or desert being flight. Nor have the great migrations come to an end today. But the numbers of mankind have increased. Moreover, Man is no longer a passive observer, and his activity, deliberately or inadvertently, contributes to the transformation of the globe. The energy needs of technological Man are today satisfied by exothermic transformation of nonrenewable resources (coal, petroleum, to some extent uranium); and the (provisional) centers of population are the points where this energy — harnessed in passing, for a moment, by Man — is released as heat to the atmosphere, together with some of the products of combustion. Atmospheric carbon dioxide is increasing rapidly, and this modification of the atmosphere may heat it. The Earth evolves under the influence of a new — this time internal — solidarity, as its inhabitants continually modify the boundary conditions of terrestrial meteorology and climate. Man-Prometheus has written on the clouds, and the brilliant Mene Mene, Tekel Upharsin is there in the sky for all to see, but Man-Belshazzar feasts on: and yet it is *we humans* who are threatened with catastrophe — not the Earth, which has seen worse, as when the dinosaurs perished!

To see the Earth and its climate as dependent on not one but a multiplicity of factors, at the nexus of internal and external influences, this is a rather new way of looking at our planet: we have come a long way since the poet sang: "Live on, cold Nature, unceasingly reborn!" Robert Kandel's book puts us at the core of this universal dependence. Focussed on the Earth, it does not confine itself to setting out the Earth's place in the Cosmos, life's place on Earth; it also and especially emphasizes their mutual connections (and should not any book, be it focussed on star or galaxy, on Man — or the termite — on atom and photon — look beyond its focus?). Pascal placed Man midway between the infinitesimal and the infinite, but even he failed to envisage the web of connections linking microcosm and macrocosm — and Man — to which this book directs our attention.

All the more reason for taking the last chapter of the book to heart. It speaks of the future of humanity, and notably of the problem of the population explosion, and its consequences. Clearly, the "models" that can be constructed for this remain imperfect. Nevertheless, to evaluate and foresee pollution, *per capita* income, nutritional deficiency, even on a global mean basis (which can hide the necessity for rapid action from the rich countries), remains significant. Political choices will have to be made: draconic population policy (democratic? Not necessarily, if there is no other way); energy production (nuclear? Dangerous, . . . but fossil fuels may pollute more); reduction of consumption in the rich countries (imperative — and thus no doubt again not very democratic?). The implications of the study of terrestrial physics must be taken into consideration, and cannot simply be shrugged off. Today's oil crisis is but a pale omen of what may be in store for us; and the oil producers should not think that they will be spared! Despite the diversity of cultures, the great variety in resources and living standards, the extreme range of outlooks, all the peoples of the Earth are incredibly bound together in this adventure. . . .

Clearly Robert Kandel does not want to play the role of Cassandra: against fear, he proposes hope — solar energy, perhaps captured in space, extraterrestrial industrial suburbs . . . Science fiction? Today maybe, but the time scales are short. Can the necessary solutions to Man's problems be found in Man's technology, based on scientific progress? Can such solutions be applied through political structures in a way which is not simply a rape of the will of the people? A host of questions have been raised, but not answered, as we reach the end of this lucid itinerary. We owe a debt of gratitude to Robert Kandel for having so clearly set down the basic principles from which must be developed a far-reaching revision of the tendencies of contemporary civilization. As Kandel writes (and as he so powerfully has shown) the history of the Galaxy commands our own history; but life itself has transformed the Earth, and Man to a large extent has become its often stubborn, always despotic, sometimes imbecilic master. Between a simplistic "ecology" which preaches a drastic reduction of consumption, but runs up against the conscious or unconscious revolt of our habits, and triumphalist technology which creates problems and then proposes still further "fixes", a third way must exist. Robert Kandel does not pretend to be able to conclude, much less to have found a "best" solution. At least he has given us the elements of a debate which will occupy us for some years yet — if we have the time.

But a preface is only an invitation to the reader; and this writer should not abuse his privilege and conclude before the reader has examined the author's statement. However, before yielding to Robert Kandel's competent and talented pen, let me express the hope that this book be read not only by students and scientists interested in understanding the mutual interactions of the diverse components of a gigantic all, but also by those who must make economic and political decisions; we would like to be sure that they always understand the nature of the problems that they are mandated to solve. And given this point, let all be conscious of the necessity of scientific research, but also of its limits, in this domain. The weight of psychological factors may be as much an obstacle to possible solutions as are the finite petroleum reserves or the underdevelopment of food production. It is no doubt out of the meeting — hopefully not too violent — of two types of sensibility, the one rational, the other quasi-mystical, that the world of tomorrow, whose outline we barely can distinguish among the multiple possibilities of today, will be born.

Jean-Claude PECKER
Membre de l'Institut
Professeur au Collège de France

Introductory Note by the author

IN writing this book my principal aim has been to present a broad overall view of the many connections between the environment of Man on Earth and the environment of the Earth in the Cosmos. Although I have tried to make this work accessible to people interested but not trained in the sciences, it was out of the question for me to write an introduction at the most elementary level to each of the fields touched on here. There are plenty of excellent books on astronomy, or physics, or biology, etc., as such, for nonscientists, or for scientist nonspecialists, and the reader who wishes to learn more about one of these branches of science, taken by itself, has a wide choice. I give a few suggestions for such reading in the Bibliography. In *this* book, the emphasis is on what connects different disciplines, and the reader, whatever his or her specialty, will necessarily encounter new concepts, and on many occasions, new language. I have tried to keep technical jargon to a minimum, and when I have used technical terms, I hope that I have remembered to define them, implicitly if not explicitly. Actually the difficulty is not so much with the rather formidable-looking terms of biology or geology, built up from Greek or Latin elements; these announce themselves as technical, and their precise meanings can be found in appropriate glossaries or dictionaries. There are rather greater difficulties in the technical use of apparently ordinary words, as is common in physics or astronomy. The red giants, white dwarfs, and black holes of modern astrophysics are not creatures escaped from a fairy tale, and a flare on the Sun is a very specific phenomenon. Words like *work* or *force* appear in our day-to-day language, but the physicist tries to use them in a very precise way. Often the reader will have to judge from the context whether such a word is being used colloquially or technically.

Questions of mathematical notation, physical terminology and units are dealt with briefly in Appendix I. Approximate values of specific mathematical or physical constants are given there or in the text only when it appears necessary to a better understanding of the material. However, I have tabulated some characteristic scales of our terrestrial and cosmic environment in Appendix II, and parameters of the different planets of the solar system are compared in Appendix III.

This book is based in part on a course for nonspecialist undergraduate students, which I developed at Boston University in 1971/72 and taught there until I left in 1974. I owe much to my students' comments and reactions, as well as to my colleagues in Boston, the two Cambridges, Paris, Meudon, Verrières and elsewhere, with whom I have discussed the questions studied here. I am indebted to the members of Section VII of the Comité National de la Recherche Scientifique for having relieved me of responsibilities which would not have allowed me to undertake this work, and to the Conseil Scientifique de l'Observatoire de Paris for its support. Many people were kind enough to furnish the photographs with which I have illustrated this book: my thanks to C. Bertaud, T. A. Croft, P. Glaser, M. Hubrecht, P. James, A. Krogvig, M. J. Martres, P. Mein and C. Pollas, and to the Météorologie Nationale. I am grateful to Stephen H. Schneider for his encouragement at a critical point. It is a great pleasure to thank my mentor and friend, Professor Jean-Claude Pecker, for his generous support, and for contributing the Preface to this book. Finally to my wife, for taking care of me and of our daughters, to my mother, for helping when it was hardest, and to my father, for reading the manuscript so carefully, goes my deepest gratitude.

1

Overview

> Knowest thou the ordinances of the heavens?
> Canst thou establish the dominion thereof in the earth?
> Canst thou lift up thy voice to the clouds,
> That abundance of waters may cover thee?
>
> Job 38.33

WE humans are the inhabitants of a small planet, moving in orbit around a middle-sized star on the fringes of our Galaxy, itself one of many in the vastness of the universe. The Sun and the stars, like clouds and the wind, mountains and rivers, forests and meadows, are part of our environment. For ancient Man this was evident, and the Sun and stars were gods, like the wind, the clouds, the seas and the Earth itself, governing the course of events. That there should be links between the regular predictable rhythms of the celestial bodies, and the largely unpredictable vicissitudes which humans have to contend with on Earth, was obvious, even if the links remained mysterious, revealed only to an élite of astrologer-priests.

With the Copernican revolution, the old geocentric world picture was overthrown, and the Earth no longer appeared to be the focus of cosmic forces. The development of the modern scientific method encouraged analysis of simplified problems piece by piece rather than attempts at global explanations of a many-faceted complex reality. For these reasons, the study of the links between the cosmic and the terrestrial environments did not progress as much as the study of the Sun, the stars, the clouds, the seas or the solid Earth, taken separately. Indeed, because of their association with astrology, attempts to relate Earth and Cosmos were often looked on with some suspicion. Of course the overriding importance of the Sun could not be denied, nor could the Moon's influence on the tides, but the rest of the universe was often treated as irrelevant.

The universe is the subject of the study of astronomy. This science teaches us that the Earth is one of several planets — relatively cold bodies — moving in nearly circular orbits around the hot luminous Sun. It teaches us that the stars are suns and that the Sun is a star — a rather ordinary one

— one of many billions making up, together with some clouds of gas and dust, the large flattened rotating system which we call the Milky Way or Galaxy. The Galaxy is in turn a member of a larger system, the Local Cluster of galaxies, and there may be a still larger Supercluster of which this cluster is a member. The universe seems at present to consist essentially of galaxies and clusters of galaxies, and radiation; it appears to be in expansion, and a majority of astronomers believe it to have "originated" in a "big bang" some ten to twenty billion (thousand million) years ago. All of this may appear to be very remote from our environment here on Earth, but in this book I shall try to show how the cosmic environment — the universe of the astronomers — including the distant galaxies, our own Galaxy and its history, the Sun, the planets and the Moon — rules or influences our terrestrial environment.

This brings us to the thorny problem of defining what we mean by the environment of Man on Earth. Various governmental and indeed international bodies have attempted this, and the result has often been definitions so vague and all-inclusive as to be practically meaningless. Of course, ambiguity has its uses, and it is true that cultural and social structures for example are important components of our human environment. Indeed we shall see that they give us the capacity of modifying the terrestrial environment in very concrete physical ways. In this book we shall for the most part be dealing with *global* rather than *local* aspects of the *physical* and *chemical* environment, and very little with the social, cultural, political or aesthetic environment of Man. Certainly the latter are important, but as an astrophysicist I do not feel myself to be particularly qualified to comment on them, although I have not always resisted the temptation to do so here. As an astrophysicist, i.e. as a physical scientist, I do believe that these non-physical aspects of our environment are the result of the development of human societies, in the presence of and in competition with other species of life, all of which have evolved *in response to* the physical and chemical environment on and of the Earth. This is not a totally one-sided affair. As we shall see, the emergence of life led to a radical modification of the physical and chemical environment, and the development of human society is continuing this, not just locally, but on a global scale. It has often been noted that astronomy differs from other sciences in that it is a science of observation: although we can construct conceptual models and perform thought experiments on them, if necessary with the aid of computers, physical experimentation is not possible. This is no longer absolutely true, since with the advent of the space age, we have been able to carry out some experiments in the traditional sense, in the near-Earth environment. However, it is more important to note that mankind is in fact engaged in a global experiment, testing the response of the planetary environment, and specifically of the biosphere and atmosphere, to the development of agriculture and of industrial technology. Since, practically speaking, this planet is the only one we've got, it would seem wise to try to understand what we are doing. I believe that an astronomer's perspective on this can be useful.

What particular aspects of the global environment shall we discuss? We shall examine the chemical composition of the Earth, and see how it is related to cosmic processes, as well as to terrestrial ones. We shall discuss the radiation environment at the Earth's surface and above, and we shall see how both solar and terrestrial radiation interact with the atmosphere. Most of all, we shall discuss climate, because our conditions for life, and our present modes of existence, depend enormously on climate. As a phenomenon of the atmosphere and surface of the planet, climate is of particular interest to the astronomer; it is also particularly sensitive to perturbation, from within as well as from without. What do we mean by climate? We shall be discussing both global and local climate, in terms of some sort of average temperature, rainfall, wind speed and direction, as well as some indication of the typical departures of these quantities from their mean values. We shall try to define this concept of climate more precisely later on, but already it is apparent that by climate we mean some sort of average over the weather.

A word, then, about the weather. As an astronomer, I have frequently had to disclaim any responsibility for the weather, and to deny that my study of the physics of the Sun and stars particularly qualifies me to predict rain or sunshine for the coming weekend. Most astronomers have had this experience. We are not meteorologists, and we generally concern ourselves with what happens outside the Earth, and not on it. And yet the layman is not wrong to ask us these questions. We all have learned that the basic day/night and seasonal rhythms are astronomical rhythms. I shall describe these in some detail later in this book. Where climate is concerned, for example the fact that in New York or Paris, it is generally warmer in July than it is in January, the basic rhythms and north-south differences are dictated by astronomical considerations. Indeed, in some languages, the same word, for example *le temps* in French, is used to denote what in English we distinguish as time and weather. While the astronomer can predict, years in advance, the exact instant of summer solstice, or of an eclipse of the Sun, no astronomer, nor for that matter any meteorologist, would venture to predict the hottest day of the year even as little as a month in advance.

How much should we make of this difference? In the recent novel by the French writer Michel Tournier, *Les Météores,*[1] accurate prediction of the motions of the planets is conceded to science, but the view is expressed that in the atmosphere, perhaps by the breath of the Spirit, mathematical time works its way out as weather, unpredictable and indescribable by science. From the point of view of natural science, this cannot be accepted. We physicists believe that the smallest eddies in the atmosphere are governed by natural law just as is the planet Earth in its movement around the Sun. The difference is that for planetary orbits, we need only consider one or a very few overwhelmingly dominant forces, to predict motions with extremely high accuracy. For weather, which is the history of the motions of eddies in the Earth's atmosphere, many different almost equal forces compete with one another and interact, so that accurate detailed long-term prediction would require extraordinarily detailed information, generally unavailable, as well as impractically high precision in the numerical computations. Thus practical limitations on our ability to predict weather, even imperfectly, appear in some cases in a matter of hours, whereas we can compute planetary motions over centuries and even millenia before these limitations appear. Few physicists would concede that weather in *in principle* indeterminate, only that in practice the accuracy of prediction is strongly and perhaps forever limited by inadequacies in data and in computation. We physicists are mostly materialists, i.e. we deny a mind—matter duality, but this does not mean that we reduce the universe to clockwork or the mind to a computer. Rather we assert that no supernatural spirit is needed to account for the fantastic diversity and variability of nature; the laws of nature and the enormous though not infinite number of ways in which the elementary particles can interact suffice. Our material universe is rich; those who would accuse us of *scientism* are generally looking at extremely crude caricatures of our view of nature.

In what follows, we shall begin our study of the connections between the Earth and the Cosmos, by a consideration of the laws of nature themselves, and how they operate in both microcosmos and macrocosmos. At the largest spatial scale possible, we shall see how the universe as a whole defines a reference frame with respect to which motions on and of the Earth must be measured. On the temporal scale, we shall look into the evolution of the universe and of our Galaxy, and we shall find that this history determines what we are made of. We shall study the multiple connections between the Sun and the terrestrial environment. Examining the planet more and more closely, we recognize the existence of a complex "geosystem", involving land, sea, ice and air, reacting to cosmic and terrestrial phenomena. We shall find that purely global considerations can be misleading, that local and fluctuating phenomena are important factors of climate and of the energy balance of the Earth. We shall show the enormous impact of life on our planet, culminating in the rise of Man. We conclude with an evaluation of the future prospects for humanity on Earth.

There are certainly other ways in which to approach these questions, and this order reflects, basically, my own intellectual itinerary as an astronomer. I have necessarily written much, indeed most of this book as a nonspecialist, and I have tried to make it accessible to people without much scientific training. While this necessarily implies considerable simplification, and at times neglect of controversial challenges to consensus views, I have tried to indicate the degree of unanimity or of controversy attached to particular points. No one person — layman or scientist — can check everything in natural science. Very few professors of astronomy have actually worked out Kepler's laws from observations of the planets, and yet we teach them to our students. In a sense we cannot avoid depending on authority, but if the scientific enterprise is to remain healthy, such authority must always be open to challenge. We depend on the scientific community as a whole to "keep us honest". Over the years the "accepted views" on many specific points undergo considerable revision, theories are abandoned, "generally accepted truths" are overthrown. This is sometimes painful, but surprisingly often not, to the researchers whose individual work is thus superseded. It is in any case a sign of the healthy state of scientific research that our ideas do evolve. The lay reader should always keep this in mind, and not be surprised when various parts of this book turn out to be wrong.

Of course, most of the work in research is the work of specialists, operating in a narrowly defined context. However, to a large degree, such work is inspired by, and finds its application and justification in, some broader framework in which the subject is viewed — a general consensus, what the historian of science Thomas Kuhn calls a *paradigm*. For example, very much current work in astronomy, especially extragalactic astronomy, is carried on in the perspective of the "big bang" theory of the expanding universe, which allows us to interpret and interrelate a wide range of observations and well developed physical theories. Thus, the attempt to determine the mass-to-luminosity ratio of a galaxy is not simply a matter of understanding galactic physics, it is part of an attempt to determine the mean density of the universe and to find out whether the expansion will continue indefinitely, or at some point reverse itself. The framework of the expanding universe is assumed. However, there are difficulties in explaining all observations in this framework, using this paradigm;

enigmas, such as that of the quasars, remain. Most astronomers and physicists view these puzzles as intriguing if annoying points of detail, but some challenge the framework as a whole.[2] If this challenge should be fully successful, it might well constitute a scientific revolution, radically reinterpreting and in some cases totally invalidating a substantial body of work in astronomy. Scientists are human, and the reactions to such challenges are not always so dispassionate as a certain mythology of the objective detached scientist would have us believe. When such challenges cannot be ignored, they are subjected to very severe scrutiny. But while there may appear to be a community of interest of many scientists resisting such challenges, even in some cases disregarding uncomfortable arguments, it is I think misleading to talk of "Official Science" *repressing* dissenting views. That is a myth cultivated by astrologers and other charlatans who would have us confuse an open mind with an empty head.

2

Matter, Radiation, and the Basic Forces of Nature

> Lo, for your gaze, the pattern of the skies!
> What balance of the mass, what reckonings
> Divine! Here ponder too the Laws which God,
> Framing the universe, set not aside
> But made the fixed foundations of his work.
>
> Edmund Halley,[1] 1686
> (transl. L. J. Richardson)
> Ode dedicated to Newton

TO discuss the universe at all, we must assume that matter, radiation, and the laws of nature are basically the same everywhere. Nearly all the information we obtain in astronomy comes from the analysis of radiation. Humans or manmade devices have not yet left the solar system, and it is not certain that the Pioneer and Voyager space probes currently on their way out will be able to perform *in situ* analyses beyond the orbit of Neptune and successfully transmit the results back to Earth. The samples of matter (cosmic ray particles) reaching the Earth from outside are certainly of great interest, but radiation tells us much more.

We begin then with electromagnetic radiation. This includes not only visible light, but also the entire spectrum from the "hardest" γ-rays to very-long-wave radio. To describe such radiation we must talk about its quantity, direction of propagation, and quality. Depending on the experiment, it may be convenient to characterize radiation either as a propagating electromagnetic wave or as a stream of particles. Apart from polarization (about which we shall say no more), the quality may be described quantitatively in terms of wavelength or frequency of the oscillation, or in terms of the energy of the particles, called photons or quanta. For someone new to physics, this multiplicity of viewpoints may be confusing at first, but it is essential to understanding the richness of nature, and it is healthy to be able to change one's mental viewpoint frequently. The wavelength λ, frequency ν, and individual photon energy E are related by the equation

$$E = h\nu = hc/\lambda$$

where c is the speed of light, and h is Planck's constant. We should note that visible light is only a small part of the complete EMR spectrum. This should be evident from Fig. 2-1.

When we consider radiation as a stream of particles — photons — we can understand the details of its interaction with matter on the microscopic scale. Let us therefore consider the microscopic structure of matter. While our theoretical understanding of this is a relatively recent development (Mendeleyev's table of the elements appeared only a little over a century ago, in 1869), a practical mastery of many material transformations is ancient. Thus where winters were cold, Man surely recognized quite early that ice could be melted to obtain water, and that liquid water froze to form ice. The basic identity of molten and solid gold must also have been known very early, wherever gold was found or traded. With metallurgy, combustion, fermentation and many other activities, Man was engaged in chemistry, well before any understanding of atoms or molecules.

At the beginning of this century it was clear that all substances could be broken down into atoms, of which there were then known nearly 100 different types — the chemical elements — each distinct, and yet with properties showing definite order. Of course there are far more than 100 distinct substances in nature, or even in the plastics or pharmaceutical industries for that matter. The smallest entity retaining the essential properties of such a substance is the molecule, which moves about freely in the gaseous state. As examples we might give water vapor (H_2O), carbon monoxide (CO) or dioxide (CO_2), methane (CH_4) or ammonia (NH_3). A molecule is a structure of 2, 3, 4, . . . or very many atoms (for the DNA molecule, the formula would fill many pages), arranged in a specific way as regards their order, spacings and geometry. Note that in the solid state, atoms or molecules may be organized in larger crystalline structures, and in general different arrangements will have different properties. Both graphite and diamonds involve only carbon atoms, but arranged in different structures.

How strong are molecular structures? Obviously this depends on the particular molecule under study, but the fact that we encounter molecules everywhere in our environment on Earth (for example, nearly all the nitrogen in the atmosphere exists in the form of the molecule N_2 and not

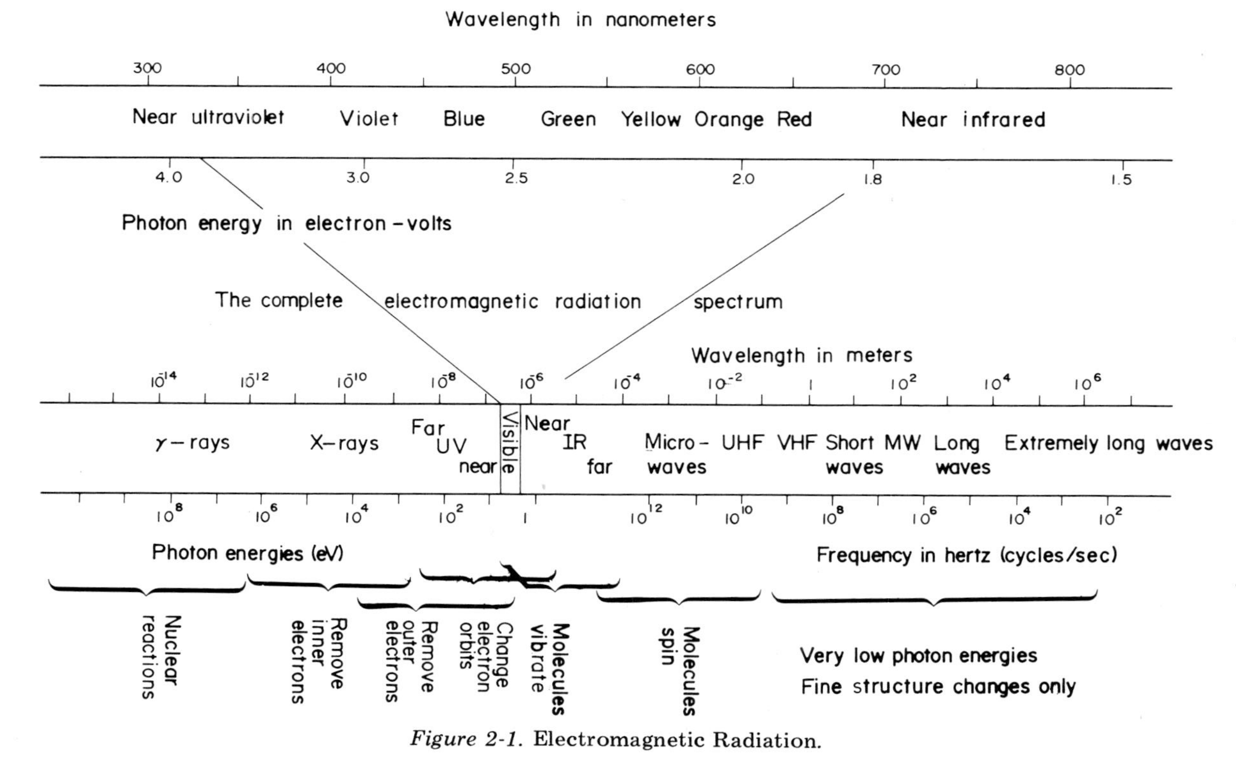

Figure 2-1. Electromagnetic Radiation.

as free N atoms), shows that the forces holding them together are reasonably strong. However, they are not so strong as to prevent human manipulation even with primitive means. Although in practical chemistry, molecules are seldom simply broken apart, one molecular structure is rather readily exchanged for another, as in the smelting of iron. To break up a single CO molecule, an energy of about 11 electron-volts (abbreviated eV) is required. Such an energy corresponds to an ultraviolet photon, or to a typical collision with another atom or molecule at a temperature of about 50 000°K, or to a moderately violent but not extremely rare collision at a temperature of 5000°K, typical of the Sun's outer layers. At the surface of the Earth, where the temperature is near 300°K and there are few ultraviolet photons around, CO survives, while on the Sun it does not. However, on the Earth it can react rather easily with oxygen to form CO_2, and not much of it survives here either.

Molecules are structures that can both spin and vibrate, and their state of rotation or vibration can be changed as a result of interactions either with other molecules or with atoms, or with photons. Rather low energies are sufficient to get molecules to spin, corresponding to radiation of wavelengths from 0.1 to 100 mm, in the far infrared and the microwave radio domains. Collisions at terrestrial temperatures or even lower also excite molecular rotation. Consequently the Earth's atmosphere both emits and absorbs such photons, and so it is both luminous and not very transparent at these wavelengths.

Molecular vibrations require higher energies, from say 0.1 to several electron-volts, the upper limit defining the dissociation energy at which the molecule is broken apart. At "room temperature" such vibrations are hardly excited at all. The state of vibration of a molecule can however be changed by interaction with infrared photons, in general. Molecules like CO, CO_2 or H_2O both emit and absorb such photons very effectively. However, molecules made up of two identical atoms, like N_2 and O_2, which constitute most of the Earth's atmosphere, cannot do so. This aspect of the interaction of molecular vibration with radiation, which is stated here without explanation, is critical to the relative transparency of the Earth's atmosphere to near infrared radiation.

We have described some ways in which molecular structures can interact with radiation, without worrying about the structure of the atoms that compose them. To understand why the elements have their particular properties, as well as to understand a host of other phenomena in nature, we must inquire into the structure of the atoms themselves. The 100-odd elements appear to be built up from three different types of "elementary" particles, two of which, the proton and the electron, have equal but

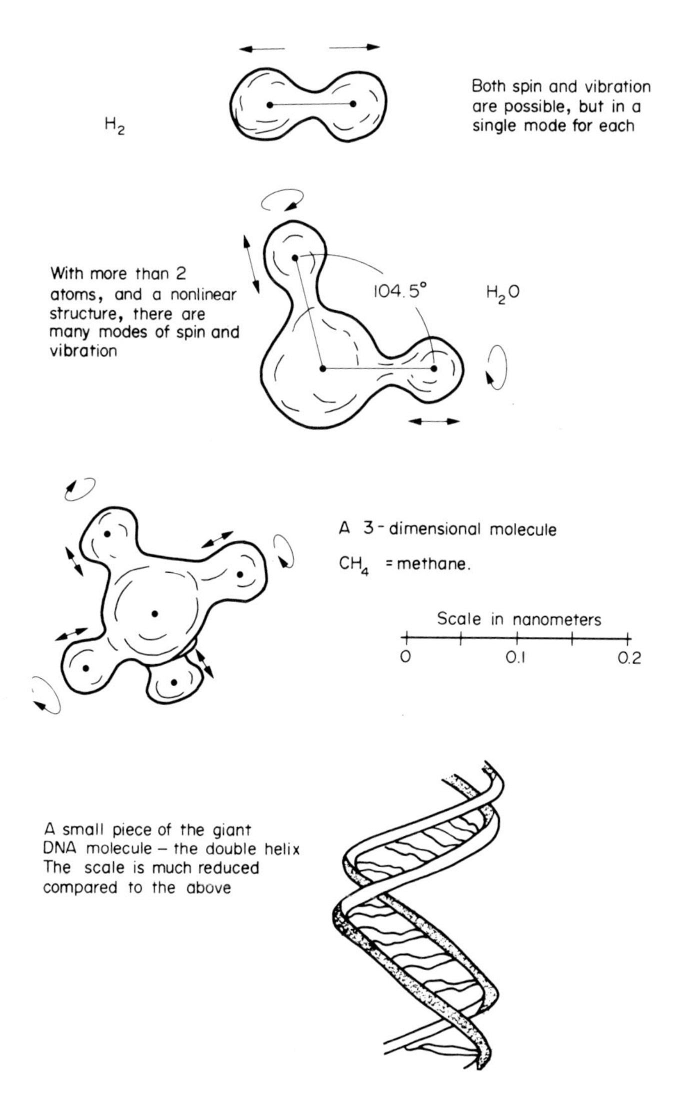

Figure 2-2. Some molecular structures.

opposite electric charge, the third, the neutron, having no charge. The neutron and proton have essentially the same mass, about 1840 times the mass of the electron. Protons and neutrons are grouped together in an extremely compact structure, the nucleus, which contains within about a millionth of a nanometer (i.e. within 10^{-15} meters) practically all of the mass of the atom, and a definite positive charge. It is this charge of the nucleus, i.e. the atomic number Z equal to the number of protons it contains, which determines the chemical properties of the atom. For the neutral atom, the same number Z of electrons are to be found orbiting the nucleus, at distances ranging from 0.05 to perhaps tens of nanometers, much larger than the nucleus itself. When one or more electrons are removed from (or added to) the neutral atom, we have what is called an ion, with a net positive (or negative) charge. Such changes, or less drastic modifications of the orbits of the electrons, can come about as the result of collisions with other material particles, or during the absorption or emission of photons. For outer electrons of atoms, energies typically of a few electron-volts, corresponding to visible light photons, may be sufficient to change electron orbits, but the energies required for first ionization — the removal of the outermost electron — correspond to ultraviolet photons. For electrons closer to the nucleus, still higher energies, corresponding to X-rays, are involved. These changes in the orbits of electrons can also occur in molecules. At any rate, since higher energy photons can usually interact with atoms, it is hard to find materials transparent to them. To do astronomy at short wavelengths, we have to go above the atmosphere.

In all of these phenomena of the microcosmos, common sense is a poor guide. We have described electrons as particles, and yet in some experiments they are more easily understood in terms of waves. One might expect that electrons could be found in orbits of any size around the nucleus, just as satellites can be put into arbitrarily distant orbits around the Earth, but this is not true. There are quantum laws, and only specific orbits are possible. Thus, as shown in Fig. 2-3, the electron of the hydrogen atom must orbit at a distance given by $n^2 a_0$, where n = 1, 2, 3, etc., and a_0 is the Bohr radius, about 0.05 nm. An orbit at a radius of $2a_0$ is just not permitted. The quantum nature of the microcosmos allows us to understand the phenomena of line and band spectra. The absorption and emission of radiation by atoms and molecules takes place at specific energies (wavelengths, frequencies) characteristic of the atom, ion or molecule involved. Each of these discrete spectral lines corresponds to a transition between two discrete allowed states (Fig. 2-4). These states may be electron orbits for atoms or molecules, or states of rotation and vibration where molecules

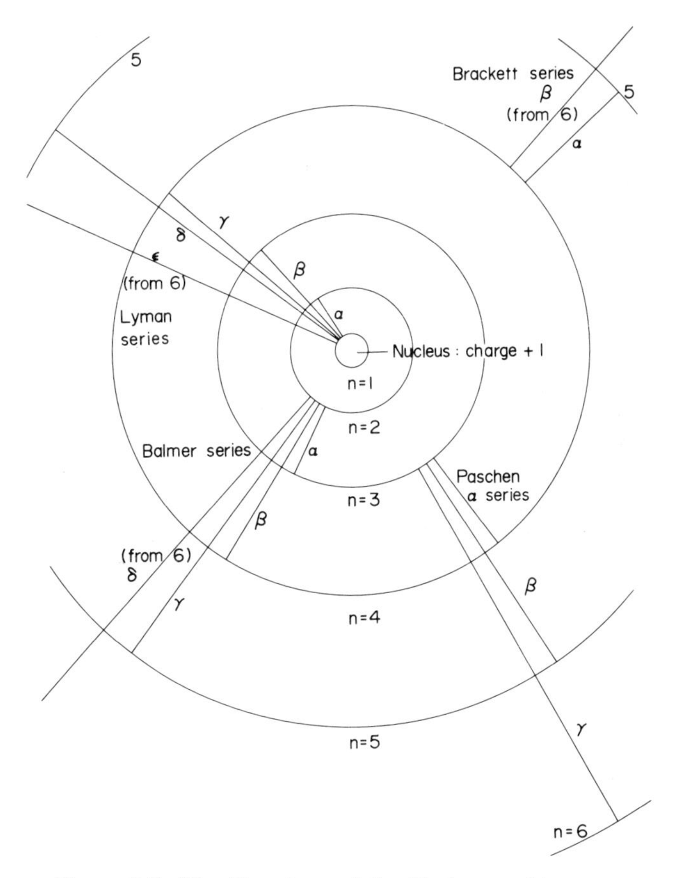

Figure 2-3. The Structure of the Hydrogen Atom as determined by the Electromagnetic Force and by the Quantum Rules. The radius of the orbit n = 1 is 0.053 nm. The first four series of transitions are indicated. Circular orbits 1 through 6 are shown to scale.

are concerned. This tool of spectroscopic analysis enables us to determine the chemical composition of the Sun and stars, an enterprise which the 19th century French positivist philosopher Auguste Comte pronounced impossible, only a few years before the discoveries of the German physicist Gustav Kirchhoff made it feasible.

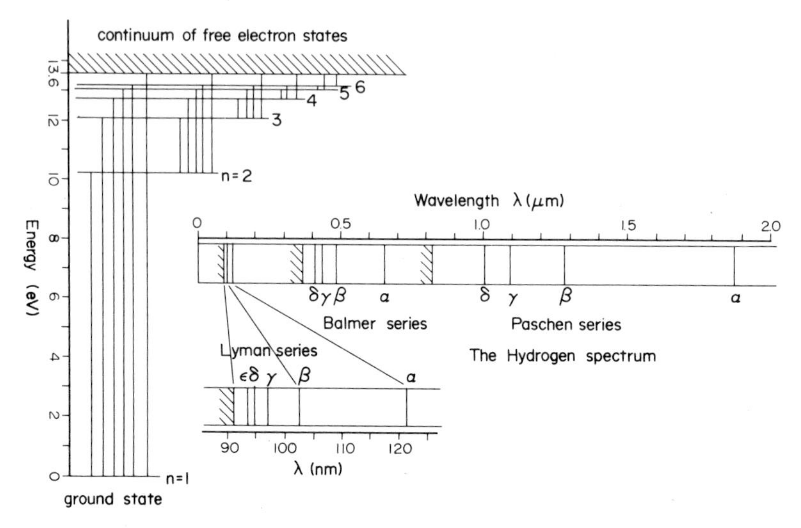

Figure 2-4. The energy levels of the Hydrogen Atom and some spectral lines corresponding to various transitions. Photoionization/recombination radiation corresponds to transitions between the discrete bound energy levels on the one hand, and the continuum of free electron energy levels on the other.

We have not reached the end of the story of matter, but what we have seen illustrates the workings of one of the four basic forces of nature, the electromagnetic force. This is the force involved in the interaction of matter and electromagnetic radiation, and it is the force which dominates the structure of matter from our scale down to the scale of atoms. The electromagnetic interaction accounts for both magnetic and electrical phenomena, and we can hardly discuss it in detail here, although we shall illustrate some of its workings on the terrestrial scale later. Here, let us consider the interaction between two stationary electric charges; this gives a force proportional to the product of the charges and inversely proportional to the square of the distance separating them, which is repulsive when the two charges have the same sign (++ or −−), and attractive when they are opposite. Thus, at the atomic scale, negatively charged electrons are attracted by the positively charged nucleus, and kept in orbit around it. It may be less evident how the electromagnetic force holds molecules together, or determines crystalline structure, since an atom as a whole is electrically neutral. Viewed from far, an isolated atom is indeed a neutral point-like particle. However, when we get close up, at distances

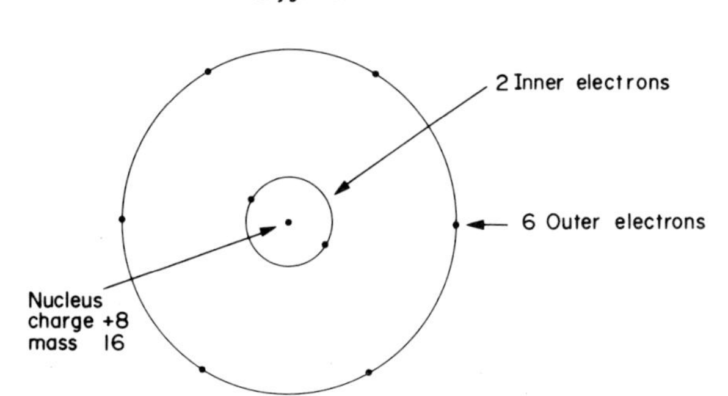

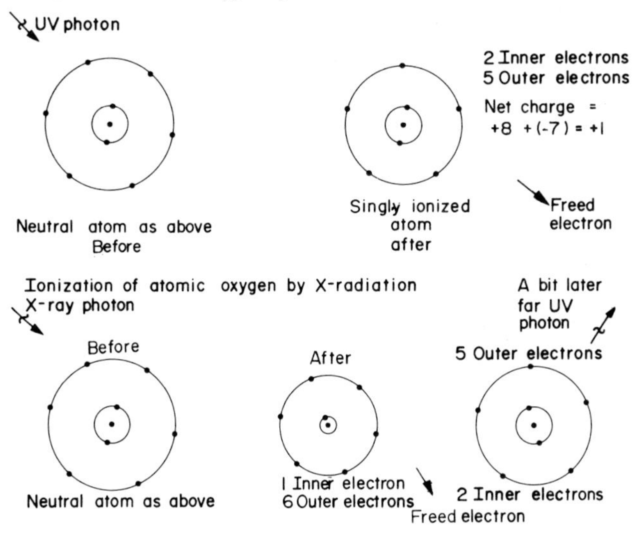

Figure 2-5. Interaction of the oxygen atom with electromagnetic radiation.

comparable to the dimensions of the electron orbits (say less than a nanometer), this is no longer true; for an approaching electron, the attraction of the atom's nucleus may not be completely counterbalanced by the repulsion of the atom's electrons, depending on which is closer. There is a residual force, which if it is attractive can lead to the formation of a larger stable edifice — a negative ion, a molecule, a crystal.

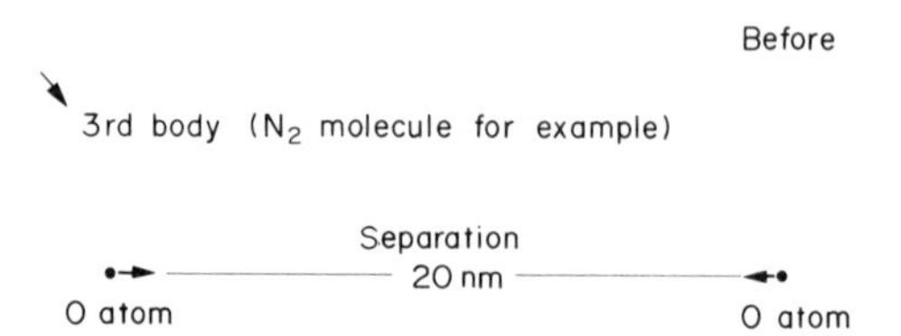

At large distances, the neutral atoms are point-like; there is virtually no electromagnetic force between them.

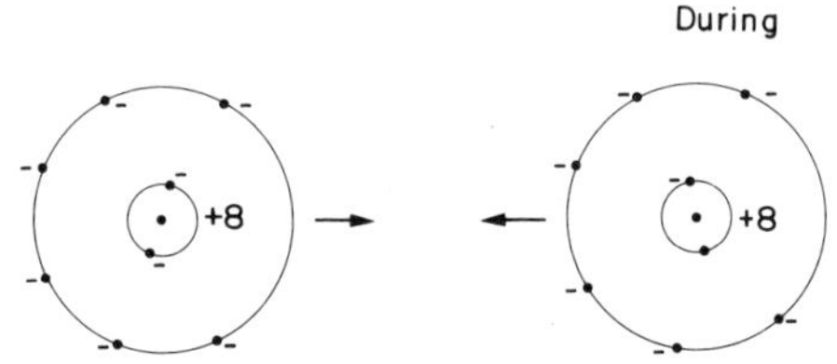

However, when their separation is comparable to their sizes, the still neutral atoms become polarized; here the one on the left appears positively charged on the right, and vice versa. There is a net electromagnetic attraction between the two.

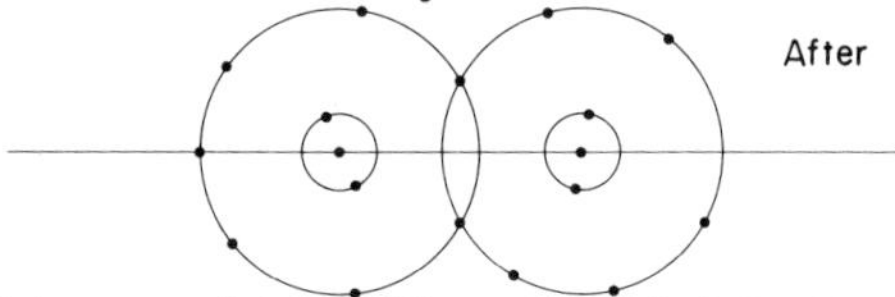

Under appropriate conditions, which include the presence of a third body, this results in the formation of an O_2 molecule; energy is released, and carried off by the third body, which may emit radiation later.

Excited N_2 molecule

Figure 2-6. Formation of molecular oxygen, by the action of the electromagnetic force.

From the scale of the atom up to the scale of mountains (with certain limitations here), whether we are discussing chemical reactions, the transparency of a gas, the hardness of a solid, specific heats, or the viscosity of a liquid, it is the electromagnetic interaction which dominates the other forces of nature. This is true even in our macroscopic world, where objects are almost always electrically neutral, with no net charge. The point is that the electromagnetic force is so strong that it usually will prevent strong net charges from surviving very long. Although our local environment is the product of all the fundamental forces of nature, it is principally the theatre of the electromagnetic force.

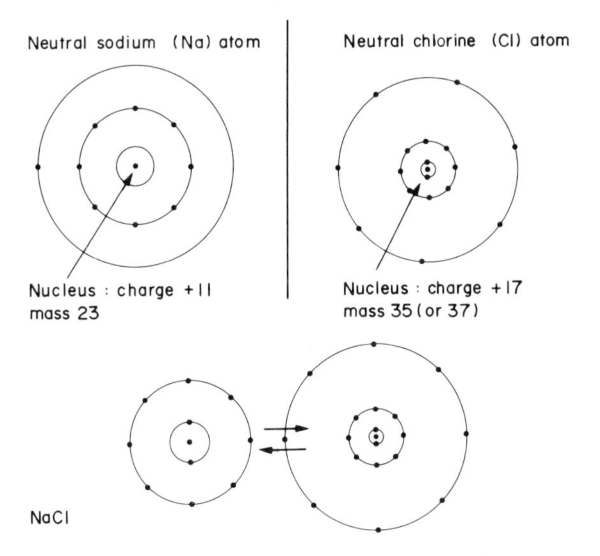

The sodium chloride (salt) molecule, when it exists

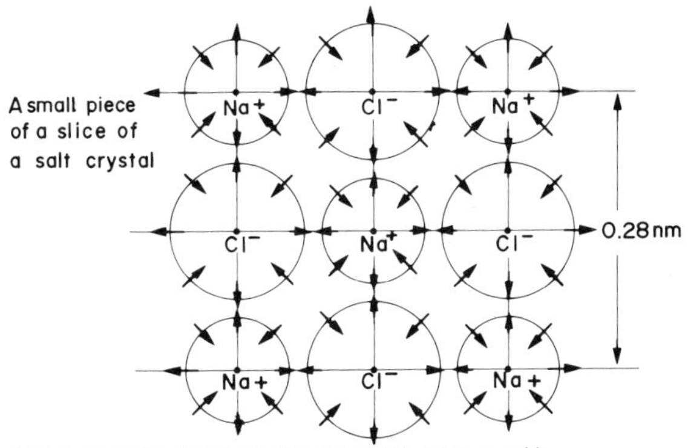

Salt is generally found in crystals which have a cubic structure, made up of positive sodium (Na^+) and negative chlorine (Cl^-) ions, held together by the electromagnetic force. The identification of an individual NaCl molecule in this structure is arbitrary.

Figure 2-7. Crystal structure.

This assertion may surprise the reader. What of gravitation? Gravitation is indeed one of the fundamental forces of nature, and it was reasonably well understood or described long before we knew much about the electromagnetic force. For much of what we see, Newton's law of universal gravitation, published in 1686, is adequate. It states quite simply that all masses attract one another, with a force proportional to the product of the masses, and inversely proportional to the square of the distance between them. Inspired by the resemblance between the electrostatic and the gravitational force laws, science fiction writers have often invoked a repulsive, antigravitational force, but in nature, the gravitational interaction is always attractive. It is, however, a very weak force. If we compare the mutual gravitational attraction of the proton and electron in the hydrogen atom, to their mutual electrical attraction, we find that the electromagnetic force is some 10^{39} times stronger! Only large accumulations of mass, like the planet Earth (6 10^{24} kilograms, or about 10^{46} atoms), can exert a substantial force. Even so, the fact that we can stand and jump illustrates that electromagnetic forces working within a few tens of kilograms of matter, can overcome, partly, the gravitational attraction of the entire Earth. It is only when matter is concentrated in an extremely small volume that gravitation can completely overcome the other forces of nature, in what is often called a "black hole". The existence of such a phenomenon was hypothesized as early as 1796, by the French astronomer and mathematician Pierre Simon Marquis de Laplace. Something like the mass of ten suns must be packed into a sphere less than 30 kilometers in radius in order to produce a black hole. Some astrophysicists believe that this can come about as a final result of the implosion and explosion processes in supernovae, about which more shall be said in Chapter 4. Certain types of X-ray sources are said to be associated with such objects. However, this is far from being generally accepted. Modern work on gravitational theory has certainly been stimulated by the prospect of finding objects such as these, for which Einstein's general theory of relativity (dating from 1915) makes predictions significantly different from Newton's laws. In recent years, the picture has been refined by Stephen Hawking of the University of Cambridge.[2] He has shown that when quantum effects are considered, particles or radiation *can* be emitted; black holes are no longer perfectly black. However, this is significant only for "mini" holes, of masses much smaller than the mass of the Earth, and comparable in size to protons. Whether such objects really exist, or are simply artefacts of a still incomplete theory, remains a matter of debate.

With the electromagnetic and gravitational forces we can understand much of what happens in the universe, but by no means all. We noted earlier that the nuclei of atoms are composed of protons and neutrons. The electromagnetic force between the protons is one of strong mutual

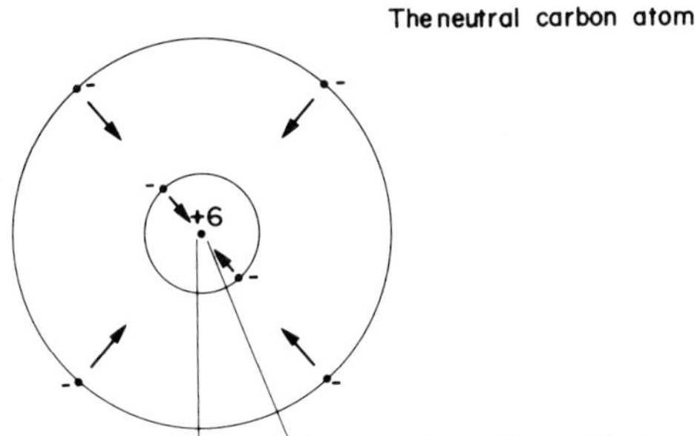

The attractive electromagnetic forces between the positively charged nucleus and the negatively charged electrons account very well for the structure of the carbon atom. Mutual repulsion of the electrons affects the structure, but the central attraction is stronger

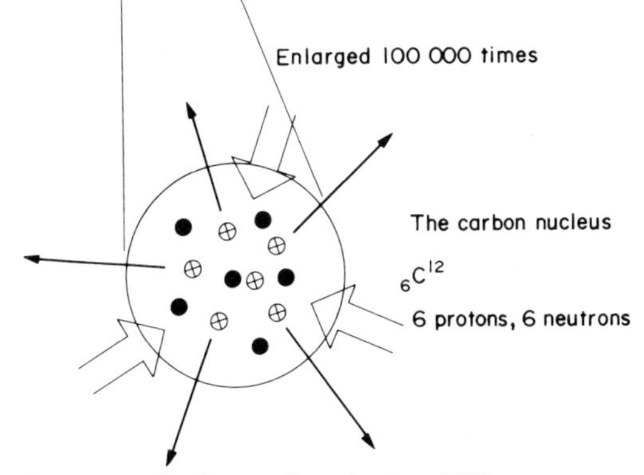

Considering electromagnetic forces, there is very strong mutual repulsion between the 6 protons, and no force on the 6 neutrons. To account for the stability of the structure of the carbon nucleus, there must be a strong nuclear force of attraction among these particles, operating over very short ranges

Figure 2-8. The Nuclear Force(s).

repulsion. The fact that the protons and neutrons remain confined in the extremely compact nucleus is evidence for the existence of a still stronger force holding the nucleus together. On the other hand, there is no evidence for the existence of this force on the atomic scale (about a tenth of a nanometer). Thus this strong nuclear force must be an extremely short-range force compared to the electromagnetic force, and must drop off with distance much more steeply than $1/r^2$. As this force binding the nucleus together is stronger, the energies involved are greater, and correspond to γ-ray photons, or to collisions at extremely high temperatures (many millions of degrees). Our understanding of this force is recent, having developed mostly since the 1930's. Our ability to manipulate this force and tap the enormous energies available inside the nucleus of the atom, is a major element of the present crisis of mankind. Let us not be misled: the greatest present danger to the environment of Man, the greatest menace to the survival of humanity, remains the existence of huge stockpiles of nuclear weapons.

The development of nuclear physics revealed quite rapidly the existence of a fourth force of nature — the weak nuclear interaction.[3] Only certain types of particles are affected by this interaction, but they include the electron. Like the strong nuclear force, this is an extremely short-range interaction. It is about 30 million times weaker than the strong force, but it still is stronger than the gravitational interaction. Without going into details, we shall note that the weak interaction is essential to understanding many nuclear processes, in the Sun and stars, in nuclear reactors, or in radioactivity. In particular, the particles called neutrinos are affected *only* by the weak interaction, unlike photons, which they resemble in other aspects since they too are massless and travel at the speed of light. As physicists reached very high energies (thousands of millions of electron-volts, or GeV, and up) in their accelerators, they found many more "elementary" particles in addition to the proton, neutron and electron. While considerable order has appeared in the pattern, there is no general consensus as to where this is leading. Will all these particles be shown to be made up of variously flavored and colored quarks? Do we need a fifth super-strong force in nature to understand them? The picture is not at all clear, and for nonspecialists, the concept of "strangeness" seems apt.

Another discovery of modern physics, related to the preceding, has been the existence of symmetries — to each particle corresponds an anti-particle. Thus to the proton corresponds the antiproton of the same mass but negative electric charge, to the electron the positron; to the normal hydrogen atom, an atom called positronium consisting of a positron orbiting an antiproton. Our local environment, including not only the solar system but also almost certainly the entire Galaxy, mainly consists of ordinary matter. The encounter of matter and antimatter leads to annihilation of the mass, with release of energy following the famous law $E = mc^2$ in the form of photons and neutrinos of high energy. We could hardly expect to find matter and antimatter well mixed here in our "quiet" environment. If, however, the two kinds of matter exist in equal amounts in the universe as a whole, as many theoretical physicists would prefer to believe (for what are basically aesthetic reasons), we would like to understand how the separation was accomplished. Some astronomers have suggested that matter—antimatter annihilation is taking place in intergalactic space and accounts for a general background of γ-rays, but the observations — which can only be made from space — remain uncertain.

Finally, we may ask how the four (or five, or . . .) basic forces of nature are related to one another. For much of his life, Albert Einstein sought a unified theory encompassing these different interactions, and many physicists are continuing the quest.[4] Progress has been made in recent years, at least in relating the weak nuclear interaction to the electromagnetic force. Much remains to be done before this physicists' dream, of encompassing all the forces of nature in a single theoretical statement, can be realized, if ever it can.

3

The Earth and the Universe

> When, accordingly, we say that a body preserves unchanged its direction and velocity *in space*, our assertion is nothing more or less than an abbreviated reference *to the entire universe.*
>
> Ernst Mach[1]
> (transl. T. J. McCormack)

HOW is our environment here related to the structure of the universe as a whole? First of all, we are linked through the fundamental laws of nature, which govern us as they do the Cosmos. We have examined these in the preceding chapter. Does the relation go further? Some physicists have speculated that it does, that various pure numbers (i.e. ratios) involved in the laws of nature are related to numbers defined by the size, mass or age of the universe. For example, the English theorist Sir Arthur Stanley Eddington speculated that the number of protons in the universe was equal to the square of the ratio of the electromagnetic to the gravitational force in the hydrogen atom, a large (10^{78}) but not infinite number.[2] On the other hand, P.A.M. Dirac of Oxford reasoned that in an expanding universe, this ratio must be changing, with gravitation progressively becoming weaker relative to the electromagnetic force as the universe ages. This view has been and still is being pursued by many other physicists, who have looked into the consequences for the evolution of the Sun and the Earth as well as for clusters of galaxies. Since the size of a body generally depends on the balance reached between the gravitational forces of attraction tending to contract the body, and the resistance of the material — essentially by electromagnetic forces — to further compression, a weakening of the gravitational force should lead to expansion of such bodies. There are a very few geophysicists who have argued that this can account for the existence of continents on the Earth, but this seems highly unlikely to most. In a similar vein, it has been argued that many puzzles in astronomy and cosmology are in fact easier to resolve if gravitation has indeed weakened during the several billion years since the "big bang".[3] However, most of the evidence now available seems to argue against present variation with time of the "constants" of nature, in particular the gravitational constant. At the same time, most astronomers believe that the evidence shows that the universe as a whole is evolving, and not in a steady state. The combination is not philosophically agreeable to all cosmologists.

Still other relations between macrocosmos and microcosmos have been proposed. On the microscopic scale an "arrow of time" seldom appears in the laws of nature, and examining a particular atomic collision for example, the motions could perfectly well be reversed. However, when we consider a macroscopic collection of atoms, this is no longer true; as described by the Second Law of Thermodynamics, there is an arrow of time, and entropy (a measure of disorder) must increase. For example, if we pour hot water into a tub full of cold water, we find it all uniformly lukewarm after a time, even without stirring. We certainly do not find boiling water some time later at one end of the tub, and freezing water at the other end, even though this would not violate the First Law of conservation of energy. How is it that an arrow of time appears in the behavior of the tub of water, when it does not apply to the motions of individual water molecules as they collide and exchange their energy? Some physicists believe that ultimately the arrow of time on this scale is determined by the expansion of the universe which defines an arrow of time on the cosmic scale. All these questions remain extremely controversial.[4]

How else can we relate our environment to the universe as a whole? Let us consider the radiation reaching the Earth. Most of this comes from the Sun; at night the sky is dark. Is this a surprise? An old argument, called Olbers' Paradox, shows that in any case it tells us something about the structure of the universe. The argument, very concisely stated, goes as follows. If the universe is infinite in extent, and uniform, i.e. everywhere and always containing similar stars at constant density, then wherever we look, our line of sight must eventually encounter a star. But then, the whole sky should be as bright as the disk of the Sun (presumed to be a typical star), and we should all be cooking at a very high temperature! What the paradox has shown us is that a particular model of the universe is wrong. For one thing the star density is high here in the Galaxy, but very low if not zero between the galaxies. The galaxies themselves are clustered, and if there is an infinite hierarchy of clustering, the night sky can indeed be perfectly dark. There are other ways out of the paradox, linked to the finite lifetimes of stars and the vast distances separating them.[5] The most commonly cited argument refers to the fact that the light from distant galaxies is redshifted, according to Hubble's law stating that the redshift increases linearly with the distance. Whether this is to be explained by the

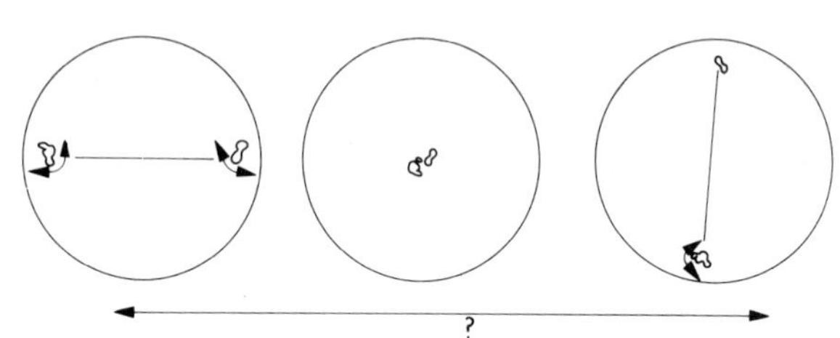

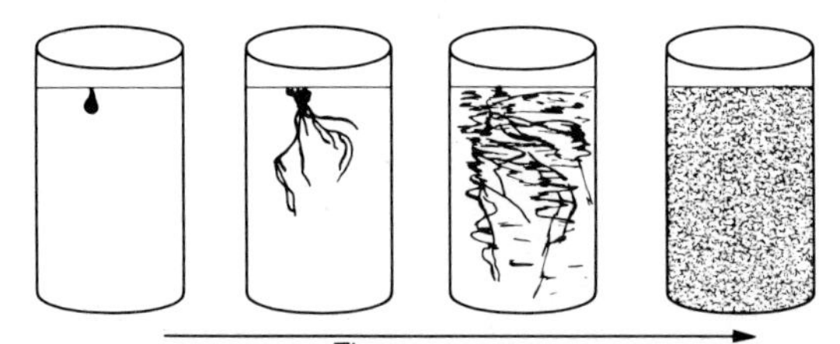

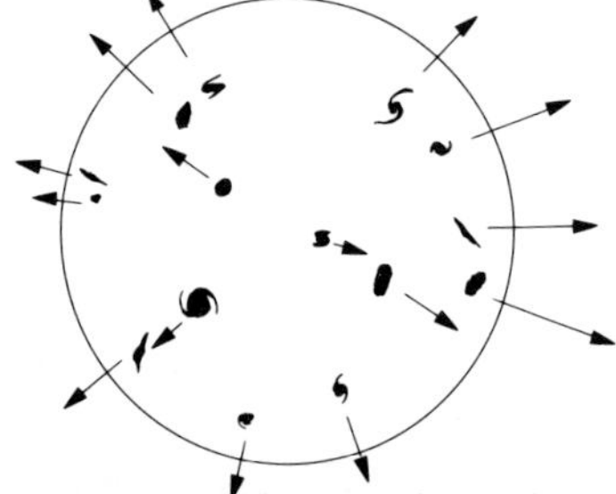

Figure 3-1. The arrow of time — a connection between cosmic evolution and phenomena on our scale?

expansion of the universe, or by some other phenomenon, makes no difference here. The sky is darkened.

In another sense, however, the sky is *not* dark at night. In the radio region of the spectrum, wherever we look, there is a nearly perfectly isotropic background intensity, corresponding to radiation from a "black body" at a temperature of about 3°K, i.e. 3 degrees above absolute zero. Some such background was predicted on theoretical grounds, as a consequence of the "big bang", by the Russian-American physicist George Gamow, and independently in Princeton by Jim Peebles working with Robert Dicke. This was a number of years before Arno Penzias and Robert Wilson (recent Nobel Prize winners in physics) first observed it in the microwave radio domain, at Bell Telephone Laboratories in New Jersey. Most, but not all, cosmologists interpret the observations as extremely strong evidence in support of the big bang theory. In this view, this is the farthest and earliest we can "see" in our expanding universe of finite age. The microwave background thus corresponds to the last opaque phase of the primeval fireball, and the radiation, probably originally in the visible and ultraviolet regions of the spectrum, has been redshifted by the expansion all the way to centimeter wavelengths.[6] Note that here we are using the term "redshift" to mean a shift to longer wavelength in general; when the shift is small, a spectral line in the blue is indeed shifted toward the red; but here the "redshifts" are so large that they take even ultraviolet radiation well beyond the red into the radio region of the spectrum. Obviously we could not be discussing these questions at a time when this radiation was still in the infrared; it was then too hot! One may ask whether there are any other survivors of the big bang, apart from these photons, and hydrogen and helium atoms. Stephen Hawking has speculated that many very small black holes may also have been formed in the primeval fireball, and that some may still be around today. However, γ-ray observations have not confirmed their existence. Since Hawking estimates that a proton-sized black hole of mass 10^9 tons would release energy at a rate of 6000 megawatts, an encounter with such an object would be a spectacular reminder of the fireworks that accompanied the birth of our universe!

When we consider the exact shape of the Earth, we find another connection with the universe as a whole. As a first approximation, the Earth is spherical, as a result of the equilibrium between gravitation and pressure. However, it is not a perfect sphere: it is slightly flattened at the poles, and bulges at the equator, so that its diameter there is greater by 21 kilometers than through the poles — a relative difference of one part in 297. For Jupiter and Saturn, which rotate more rapidly than the Earth even

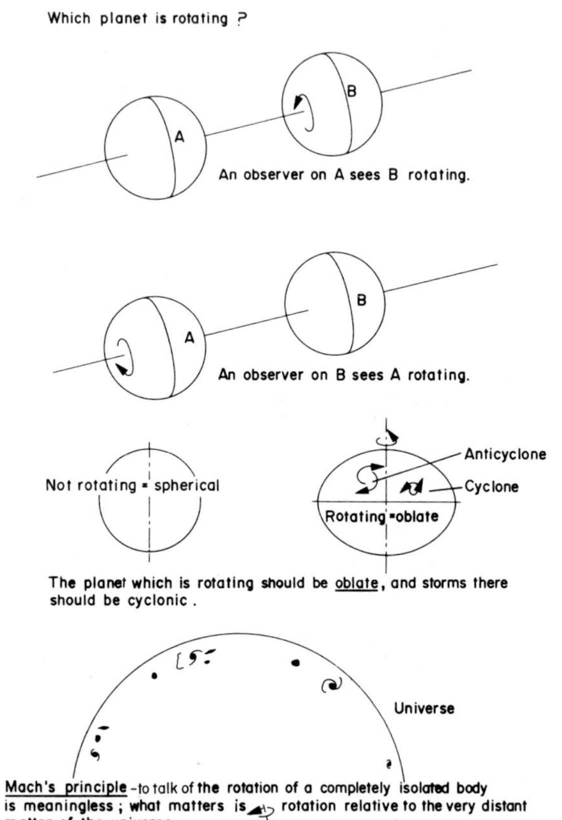

Figure 3-2. Planetary rotation, the inertial frame, and the Universe.

though they are much larger, the oblateness is much higher, easily detected on photographs. We can explain such departures from sphericity as the result of the centrifugal force associated with rotation. Another force — the Coriolis force — arises when bodies are in motion relative to a rotating planet, and we shall see that this force is essential to understanding the circulation of the atmosphere, and determines the flow in cyclonic disturbances, as we shall see in Chapters 9 and 10.

The question arises — in what frame of reference is rotation to be measured? The ancients thought that the skies turned around the Earth, and we are taught that the Earth turns. We call the appropriate coordinate system an inertial frame of reference, but that is just giving it a name, and does not tell us how to find it. Newton's laws of motion tell us how objects move, provided that we observe them in such an inertial frame, so that indeed we must define it if we wish to understand the motions of the planets around the Sun, under the influence of gravity; Newton's law of universal gravitation is not enough in itself. Similarly, we can account for the orbits of stars in our Galaxy and others, as the result of gravitation operating in an inertial frame, which then must be determined on a scale large compared to galaxies. Newton, early in his *Principia*, after remarking that "I do not define time, space, place, and motion, as being well known to all",[7] nonetheless found it necessary to elaborate on the notion of absolute space, which would determine the inertial frame (and more), but he did admit that it cannot be seen. He then went on to argue that "absolute circular motion" *can* be detected and measured, notably through the centrifugal force. The "luminiferous aether" proposed in the 19th century as the support of electromagnetic radiation, was thought for a time to define absolute space, but its existence is denied by experiment.

In fact, the notion of absolute space was strongly resisted (notably by Bishop Berkeley)[8] from the time Newton stated it, and Einstein's General Theory of Relativity seems to have done away with it entirely. The problem of the inertial frame remains however, albeit in different form. The most promising answer to the problem appears to be given by Mach's Principle: the inertial frame, and the very fact of inertia, are determined by the overall distribution of matter in the universe. A single isolated object could have no inertia, and there would be nothing relative to which we could measure its rotation or its acceleration. The British physicist D. W. Sciama[9] has attempted to derive a specific law of inertial induction, expressing the way in which the inertia of a body depends on matter elsewhere, this force to be added to the basic forces described in Chapter 2. This approach may or may not succeed. In any event, we have found a link between the universe on the largest possible scale, and both the shape of the Earth and the flow of the wind in its atmosphere.

4

The Earth in the Galaxy

> What a wonderful and amazing scheme we have here of the magnificent vastness of the universe! So many suns, so many earths. . .
>
> Christiaan Huygens,[1] 1698
> *Cosmotheoros*

THE overall structure of the universe defines the conditions of motion on Earth and in our solar system. The early history and the age of the universe determine how dark our night sky is. The composition of the solar system, of the Earth, and of the very matter of which we are made, reflects the history of our Galaxy. We shall see how in what follows.

The results of astronomical observation — the Hubble law of expansion of the universe, estimates of the average density of the universe — combined with our knowledge, both theoretical and experimental, of nuclear and high-energy physics, allow us to calculate "models" of what took place in the early phases of the "big bang". These calculations can account for — and indeed they predicted — the brightness of the night sky, i.e. the 3°K cosmic background radiation. These models and calculations also predict a universe consisting entirely of hydrogen and helium.

Now while the universe is indeed made almost entirely of hydrogen and helium, it does include a significant amount of heavier elements. They are particularly significant to us, since we are made of them. Certainly the human body, which is mostly water, contains a good bit of hydrogen, but its mass is mostly in oxygen and carbon, and many other elements, such as nitrogen, phosphorus, sulphur, iron and magnesium, to name only a few, are absolutely essential to life. All these elements, we believe, were produced in the course of the normal evolution of stars in the early history of the Galaxy, prior to the formation of the solar system some 4700 million years ago.

Our knowledge of the composition of the Earth is based partly on direct analyses of material at or near the surface, and partly on indirect evidence regarding the interior, in particular the iron-nickel core. We also have analyzed many meteorite samples of different types, as well as a large number of samples of rocks and dust from a few sites on the Moon. The Soviet Venera probes have performed analyses in the atmosphere and for a time on the surface of Venus, while the U.S. Viking laboratory operated for an extended period of time on the surface of the planet Mars. Thus we have some "direct" analyses, of the sort that would have satisfied Auguste Comte, of the chemical composition of a few samples of extraterrestrial material. In addition to all this, we have of course analyzed the chemical composition of the sun using the spectroscopic method.

These analyses reveal both striking differences and strong similarities. Thus the Sun is mostly hydrogen and helium, while the proportion of these elements on Earth is relatively low. On the other hand, the ratios of the abundances of titanium and vanadium are about the same in these various bodies. The atmosphere of the Earth is mostly molecular nitrogen and oxygen, while that of Venus is mostly carbon dioxide, as is that of Mars, albeit at much lower pressure. Now the relative absence of hydrogen and helium on planets like the Earth, Venus and Mars is relatively easy to understand. These elements are light, and in gaseous form they escape easily from these planets, at the prevailing temperatures. When we study the more remote planets, such as Jupiter and Saturn, where temperatures are lower, we find a very high proportion of hydrogen in various forms. These planets are very massive (compared to the Earth) because they have not lost their hydrogen and helium, and they have not lost these elements both because they are so massive that escape speeds are high, and because it is and always was cold there, so that atoms and molecules move slowly. The escape factor is thus very important in accounting for differences in composition between the Earth and the Sun. It does not however account for all the differences. For example, the element neon is relatively rare on Earth compared to the Sun, and some researchers believe that it was swept away by a very strong solar "wind" (see Chapter 6) when the Sun was still a young star.

Generally speaking, we nevertheless believe that the evidence shows that the planets and the Sun were formed from the same material, and started off their histories with the same chemical composition. The differences that are now observed must be the consequences of the processes by which the planets were formed from the pre-solar cloud, and of the evolution that has taken place since then. These processes can be quite complex, many details are debated, and we certainly do not understand them all yet. Let us however take as a reasonable first approximation the hypothesis of a single chemical composition for the solar system at its birth. We can then combine the solar, terrestrial, and meteoritic composition data and obtain a table of abundances complete for all the chemical elements. This table, the standard abundance distribution, presumably gives the composition of the pre-solar nebula from which the solar system was formed.

When we examine the standard abundance distribution in detail, we find many correlations of abundance with properties of the nuclei involved. This requires that we distinguish between isotopes — nuclei having the same number of protons, but different numbers of neutrons. Examples are normal hydrogen ${}_1H^1$ whose nucleus consists simply of a single proton, contrasted with rare deuterium ${}_1H^2$ whose nucleus contains a neutron as well as a proton and thus has twice the mass. We also often encounter the isotopes of carbon — ${}_6C^{12}$, ${}_6C^{13}$, ${}_6C^{14}$. In recent political history, and today too, the isotopes of the very heavy elements uranium (${}_{92}U^{235}$ and ${}_{92}U^{238}$ especially) and plutonium, have played an important role. For each of these sets of isotopes, since the corresponding atoms have the same numbers of electrons, their chemical properties are identical. Apart from usually relatively small effects due to the mass differences, the differences between isotopes, notably in abundance, are due to pheno-

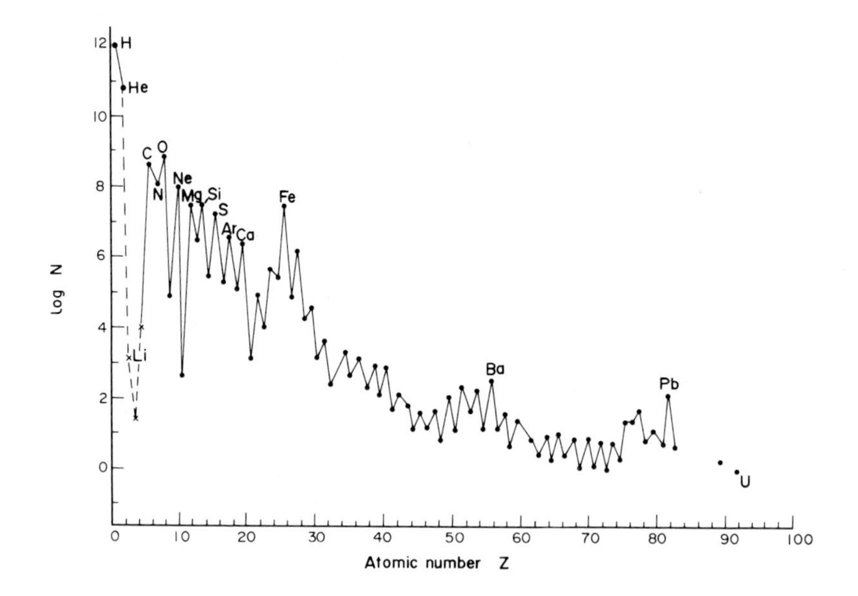

Figure 4-1. The Standard Abundance Distribution — logarithm of the relative number of nuclei as a function of Z. (after Cameron)[2]

mena at the nuclear scale, involving the nuclear forces. Returning to the standard abundance distribution, we first note the overwhelming dominance of hydrogen ($_1H^1$) and helium ($_2He^4$). Going to higher charges and masses, we note that the nuclei which can be made by combining 3 or more helium nuclei (also called α-particles) are relatively abundant — $_6C^{12}$, $_8O^{16}$, $_{10}Ne^{20}$, $_{12}Mg^{24}$, $_{14}Si^{28}$, although there is a general drop-off with increasing nuclear mass. This trend is interrupted by a peak of relatively high abundance surrounding the iron nucleus $_{26}Fe^{56}$.

The very heaviest elements are extremely rare. Some of them are not found in our solar system, outside of physics laboratories, for example Einsteinium ($_{99}Es^{252}$) or Mendelevium ($_{101}Md^{256}$). This is not surprising, since these nuclei are intensely radioactive. The balance of forces in such a nucleus is precarious. Within a finite time, the nucleus emits various particles (electrons, neutrons, α-particles, photons, neutrinos, . . .) and a nucleus of lower mass and/or different charge will result. In some cases, notoriously that of $_{92}U^{235}$, fission takes place — the nucleus splits into two roughly equal parts, releasing other particles at the same time. Not only all of the heaviest elements, but also many isotopes of lighter elements, such as $_6C^9$, $_6C^{10}$, $_6C^{11}$, $_6C^{14}$, $_6C^{15}$ and $_6C^{16}$ for carbon are radioactive. The measure of the stability of a radioactive nucleus is its half-life, which is the length of time in which half of the nuclei in a large sample would undergo decay. There is no way of predicting when an individual nucleus will decay; we can only describe the process statistically. Thus the most radioactive isotopes or elements are those with the shortest half-lives. For us to have any chance to find a nucleus with a very short (fraction of a year) half-life, we must produce it, for example in an accelerator, a reactor, or a nuclear explosion, or observe the impacts of cosmic ray particles. Actually, since the energetic particles and photons emitted during radioactive decay can do considerable damage to living cells and particularly genetic material, we are not eager (to say the least) to encounter such short-lived nuclei in quantity, except under strictly controlled circumstances. The hazards linked to the operation of accelerators are extremely local and generally well controlled. Whether the hazards of power reactor operation and nuclear waste disposal can be kept local and completely under control, is a matter of intense public debate in many countries, with important impacts on economic and political planning. As for the hazards related to the explosion of nuclear weapons, these are so well known that the major nuclear powers have agreed to forgo nuclear weapons testing in the atmosphere, and to limit underground testing, despite strong pressures from military figures to continue such testing. In any event, the stockpiles of weapons remain, and the production of weapons continues.

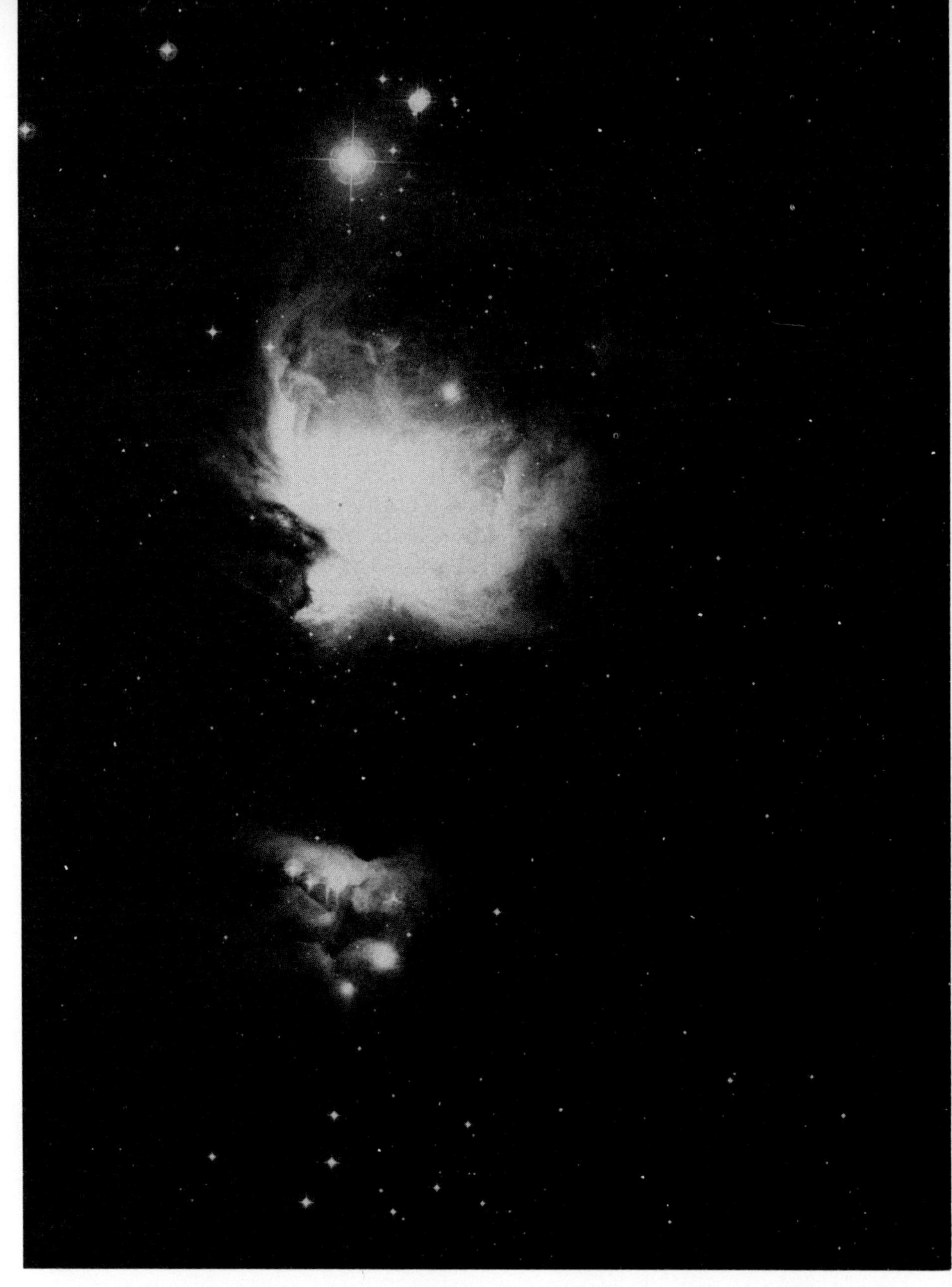

Figure 4-2. The Orion nebula, a region of star formation. Recently formed hot stars emit large numbers of ultraviolet photons which ionize the gas (mostly hydrogen); this in turn emits visible light, especially lines of the Balmer series of hydrogen (cf. Fig. 2-4), as the electrons cascade down to the ground state of energy. Note the dark areas which correspond to dense clouds of absorbing dust; it is believed that stars are still forming here. The crosses on the bright stars are an instrumental effect. (Photograph J. L. Heudier courtesy Observatoire de Calern du C.E.R.G.A.).

We have explained why the very heaviest "artificial" elements, having extremely short half-lives, are not readily found. The fact that we do find *any* radioactive elements on Earth, in particular uranium and thorium, which go through a series of decays ending up in the lead nucleus $_{82}Pb^{206}$ with an equivalent half-life of about 4.5 billion years, is proof that the Earth was not formed a very long time ago, as compared with that half-life. If the Earth were 50 billion years old, there would be virtually no uranium left. From various isotopic and elemental ratios, we can estimate the times at which various rocks were formed, i.e. since which they have not undergone significant transformations. Studies of the oldest earth rocks, of some lunar material, and of meteorites, particularly the type known as carbonaceous chondrites, all point to an age of about 4700 million years for the solar system. The radioactive elements, essentially those heavier than lead, were formed some time before then or are the result of the decay of progenitor elements formed earlier.

Let us then come back to the question of how all these elements were formed, starting with only hydrogen and helium. Calculations seem to show that the primeval fireball does not stay hot enough and dense enough for a long enough time, for the heavier nuclei to be produced. Somehow, however, out of the material dispersed in the big bang, clumps of matter formed ultimately becoming galaxies. Let us define galaxies as more or less concentrated flattened collections of gas, dust, and stars (and black holes?) containing from 10^5 to 10^{11} solar masses, typically up to a few tens of kiloparsecs in size (i.e. about 100 000 light-years). The structure of galaxies depends mostly on gravitational forces, to some extent on electromagnetic forces, especially where the dust and gas components are concerned. Somehow, in these galaxies, much more highly concentrated collections of matter form: stars, typically separated by distances much larger than their dimensions. Just how, we do not know; the observational evidence links young stars in our galaxy and present-day star formation to the densest cool clouds of dust and gas, but some galaxies seem to contain very little dust and still have young recently formed stars. In any event, at the earliest epoch when there presumably was only hydrogen and helium, there could not have been much dust, and yet stars did form.

What, basically, is a star? Suppose we define it as a collection of matter held together by gravitation, emitting electromagnetic radiation. We can understand much of its structure in terms of the balance between gravitation on the one hand, and the resistance of the gas (due to electromagnetic forces in most stars, to nuclear forces in neutron stars) to further compression on the other. Calculating the internal structure of such a body we find that generally the temperature and the density increase

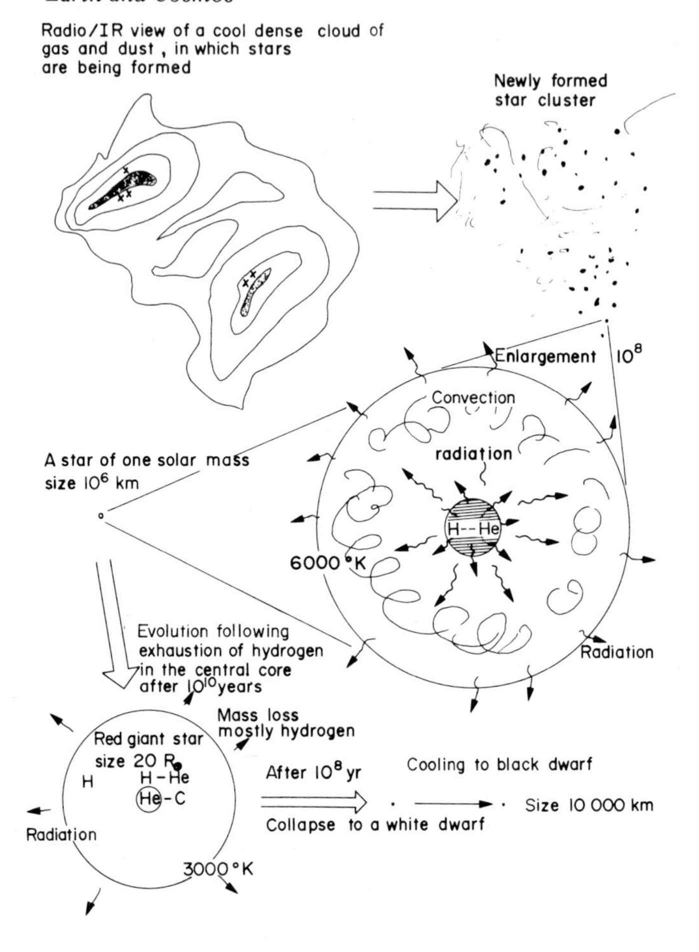

Figure 4-3. The life history of a low-mass long-lived star.

with the pressure toward the center, and that except possibly in cases of very rapid rotation or very close double star systems, the star is very nearly spherical. Now the fact that a star like the Sun radiates energy at a fairly constant rate for a long time requires explanation. Gravitational contraction is not an adequate energy source. The energy source is found in the thermonuclear reactions that take place at the high temperatures and densities reached in the centers of such stars. In the Sun, through various "fusion" reactions, 4 hydrogen nuclei (protons) are converted into a helium nucleus (an α-particle, consisting of 2 protons and 2 neutrons), with the emission of neutrinos, 2 positrons, and about 25 MeV of energy in γ-ray photons. Both the strong and weak nuclear forces are involved.

Now this may explain why the Sun shines, but it does not explain how the carbon in our bodies was formed. Indeed, stars of the Sun's mass or less, make no contribution to the production of heavy elements. When such a star, after about 10 billion years (or longer, for less massive stars), has converted all the hydrogen in its central core to helium, it evolves fairly rapidly through a "red giant" stage, during which its central regions become denser and hotter, while its outer envelope becomes enormously distended and cooler. Finally the central regions become so dense that a state called "electron degeneracy" sets in, in which the atoms are in a sense "touching" each other. The central regions can get no hotter, the energy output drops, and the outer layers collapse, for lack of support. The star becomes a "white dwarf", comparable in size to the Earth even though containing a solar mass; it then cools off quite rapidly and no longer radiates thereafter. Barring a virtually impossible collision, or substantial accretion of matter from the interstellar medium which would "reactivate" the star, the star is dead, its matter lost to the "cosmic recycling" processes of the Galaxy.

We still have not formed the carbon and iron in our bodies. It is only when we have a fairly massive star, at least 1.4 times as massive as the Sun, that interesting possibilities arise. These stars too spend most of their lifetimes in what is called the main sequence phase, during which, like the Sun, they are "burning" the hydrogen in the central core. When a massive star has exhausted its hydrogen, it too evolves to become a red giant. Now, however, in the contraction of the central regions, the temperature and density — in contrast to the situation for the less massive stars — finally reach values high enough for the "triple -α" reaction to take place: 3 helium nuclei collide, and form a carbon nucleus. This is the starting point for the production of all heavier nuclei, with the nuclei obtained by further addition of α-particles being favored. Depending on the temperature and density, and on the composition reached, other types of reactions also take place; thus we account for ${}_7N^{14}$ between ${}_6C^{12}$ and ${}_8O^{16}$. Indeed we can build up heavier nuclei all the way up to ${}_{26}Fe^{56}$ while releasing energy which accounts for the luminosity of the star. At the culmination of such an evolution, and provided little internal mixing occurs, the star might develop an "onion-skin" structure, hydrogen and helium in the outer envelope, then successively helium, carbon, and heavier elements until iron is reached in the core.

Neutrons are released in some of the nuclear reactions during these phases, and when these neutrons collide with certain nuclei, heavier elements can be built up, even beyond iron. This can account for the so-called "s-process" (for slow neutron capture) elements, such as strontium, barium, zirconium, and even in part for lead. However, the reactions building up the elements heavier than iron are endothermic; they soak up energy rather than produce any. During these phases the net energy yield from all the reactions taking place becomes smaller and smaller, while the energy outflow from the star is very high, and can become enormous when neutrinos are produced; thus the evolution of the star must proceed more and more rapidly.

Finally the energy crisis looms. Even if a star should succeed in evolving in quasi-regular fashion up to the Fe^{56} peak — which may be extremely unlikely — there would be no way for it to maintain its luminosity thereafter. Neither could we account for the existence of many of the heavier elements, without something new happening. Actually, quasi-static evolution of a star may well not get beyond the production of Ne^{20} before explosive phenomena arise. The details of the ignition and especially the hydrodynamics of such explosions are still being worked out, and may depend on the particular circumstances of the star involved — initial composition, membership in a binary system. Anyhow, certain generally agreed-upon aspects of the models of such explosions do seem to account quite well for the relative abundances observed around the iron peak. Moreover the energies and the numbers of neutrons released in such explosions do appear quite adequate to account for the production of the "r-process" (rapid neutron capture) elements, which include notably the very heavy radioactive elements such as uranium. These explosions are observed to occur in nature. They are called supernovae, and during such an explosion a star may become 100 to 1000 million times as bright intrinsically as the Sun, and remain so for several weeks. The explosion disperses most of the material of the star.

This is the most spectacular way in which material processed inside a star — "astrated", to use the word invented by the Québecois Parisian astrophysicist Hubert Reeves — can be returned to the interstellar medium of the Galaxy, and recycled, assuming that this enriched material can be condensed to form new generations of stars. It is not the only way. There is fairly good evidence that many stars lose mass at a rate which becomes significant when continued over long periods of time. Red giants and supergiants are examples. It is likely but somewhat less certain that the material lost, which comes from the outer envelope of the star, is significantly enriched in the heavy nuclei produced in the central regions. This may not

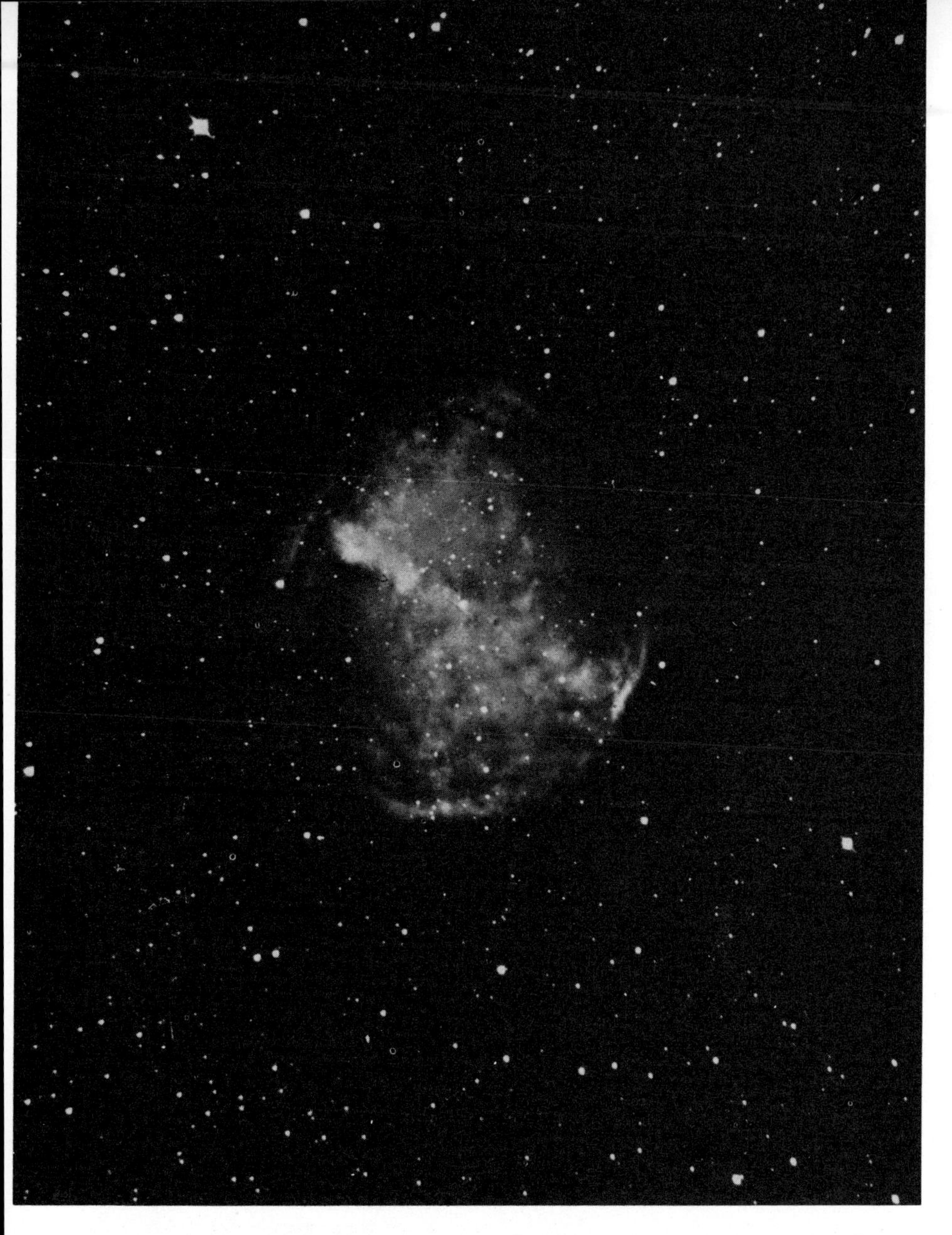

Figure 4-4. The Dumbbell Nebula (NGC 6853) in Vulpecula, example of a planetary nebula. This is one way in which a star "dies". The central star is collapsing, becoming extremely dense and hot, but it has left behind a shell of material which is ionized by its far ultraviolet photons, and then emits visible light. Some of this material will be recycled in the interstellar gas and ultimately in other stars. (Observatoire de Haute-Provence du Centre National de la Recherche Scientifique).

Figure 4-5. The Veil Nebula in Cygnus, part of a huge shell of gas expelled in a supernova explosion over 50 000 years ago, which has gradually been slowed down by the resistance of the interstellar gas. Recycling of the material is well under way. (Photo R. Chemin, courtesy Observatoire de Calern du C.E.R.G.A.).

Figure 4-6. The spiral galaxy M51, in Canes Venatici. Both hot stars and the absorbing dust lanes outline the spiral structure extremely well. There is a second smaller galaxy at the bottom of the picture, interacting gravitationally with the first. (Observatoire de Haute-Provence du Centre National de la Recherche Scientifique).

always be the case. There is however a dramatic example of enrichment of the outer layers of a star, found in the spectrum of the star FG Sagittae, which revealed very substantial increases in such s-process elements as Y, Zr, La, between 1972 and 1974.[3] This star does seem to have lost mass in the historical past, through the expulsion of a "planetary nebula", and it may do so again. Thus there is hope that it will contribute these rare metals to the interstellar medium, sometime in the "near" future.

An overall picture emerges from these considerations of the nuclear evolution of stars. Our own composition is the result of the processing of material — initially pure hydrogen and helium — in one or more generations of massive rapidly evolving stars. Through both "slow" mass loss, and the catastrophes of supernova explosions, astrated material was recycled, and the interstellar medium enriched in heavy elements. In most of the volume of the Galaxy, in the more or less spherical halo, star formation does not seem to be continuing today. Only stars having very little in the way of heavy elements in their atmospheres are found. Presumably they were formed before astration could significantly enrich the gas. The gas seems to have collapsed to the plane of the Galaxy very early in its history. Indeed, most of the mass of the Galaxy is now in the form of stars, and relatively little is left as gas. However, in the plane or disk of the Galaxy, the density of the gas has remained high enough for stars to continue to form, especially where it is increased locally by the spiral structure density wave pattern. Although enrichment seems to have been most effective during the early history of the Galaxy, it has continued albeit at a much slower rate, since the solar system was formed. All of this has contributed to making us what we are today. The carbon and oxygen in our bodies were probably formed in the core of red giant stars, sometime during the first few billion years of our Galaxy. The iron in our blood may well have been formed in the heart of a supernova, which took place before the solar system was formed some 4.7 billion years ago, perhaps as little as a million years before.[4] The uranium which tempts us as a source of energy also links us to such an explosion. Our Earth environment carries in it the history of the Galaxy.

The history of the Galaxy did not end with the formation of the solar system. How have our histories since then been linked? The distances between stars are very large compared with the scale of the solar system, typically 300 000 A.U., where 1 A.U. is the mean distance of the Earth from the Sun. Thus even a single collision of stars in the entire Galaxy, over its entire history, turns out to be a highly improbable occurrence. However, we have seen that some stellar events, and notably supernova explosions, are extremely powerful phenomena. Given the (rather skimpy)

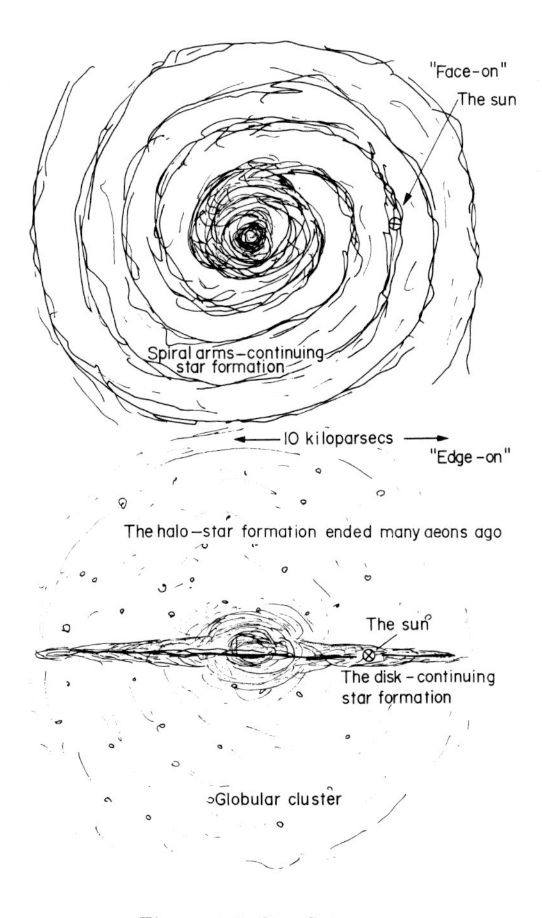

Figure 4-7. Our Galaxy

statistics on supernova occurrences, the Soviet astrophysicist I. S. Shklovsky has estimated that the solar system has several times in its history been closer than 10 parsecs (= 2 million A.U.) to a supernova explosion. The total energy flux reaching the Earth from such a supernova, unless it were very nearby indeed, would probably be much less than the solar flux, and so would have no significant direct effect in heating the Earth. However, the indirect effects, such as the changes in the structure of the Earth's atmosphere due to the ultraviolet flux (which would be much larger than the solar flux), might be quite important. We shall see how some of these indirect effects can operate in later chapters. More important, it appears that supernova remnants may be the major source of cosmic rays in the Galaxy. The cosmic rays — protons, helium and heavier nuclei, electrons — travelling at nearly the speed of light and thus carrying much energy — are probably an important factor in the mutation rate among living species. Shklovsky has suggested that cosmic rays from a nearby supernova may have been a major factor in the extinction of the dinosaurs, but this is strongly contested. However, the possibility of significant effects on biological evolution is quite real.

We have stated that the history of our Galaxy is a history of star formation continuing to this day, essentially we believe because relatively dense clouds of dust and gas persist in the disk of the Galaxy. Our solar system lies in the disk, and moves in a roughly circular orbit around the center of the Galaxy, completing the circuit in about 200 million years, which we might call a Galactic year. Thus we have completed over 20 circuits since the Earth was formed. While star collisions appear unlikely over such a period, passages through dust clouds may be relatively common. The densest dust clouds in other galaxies seem to outline the spiral arms quite well, and if we accept the idea that dust formation proceeds best where the gas density is highest, we can readily understand this in terms of the density-wave theory of spiral structure. In this theory, the stars in circular orbits around the galactic center are moving relative to the density waves, which constitute the spiral arms, and the fact that bright hot stars, like the dust, outline the spiral arms rather well, is simply a consequence of their very short lifetimes. They do not move far from their birthplaces before dying. Stars like the Sun, on the other hand, which have lifetimes of many Galactic years, must pass through the arms many times. It has been suggested that the effects of these passages appear as major climatic variations on Earth, in particular as ice ages. We shall see in later chapters that there are many different explanations for the ice ages, and that climate can be a very subtle affair, with indirect triggering effects playing a major role. It is hardly absurd to consider even small changes in the environment of the solar system as possibly modifying the climate, but dust clouds are only one of many possibilities. Moreover, despite initial intuitions, the effect of passing through a dust cloud is not necessarily to cool things off. In any case, while the density-wave theory works with smooth distributions of gas, the galactic reality involves gas and dust concentrated more or less compact and dense clouds. The spiral arms are not continuous. A statistical argument is necessary to evaluate the probability of passing through a cloud of a given density.

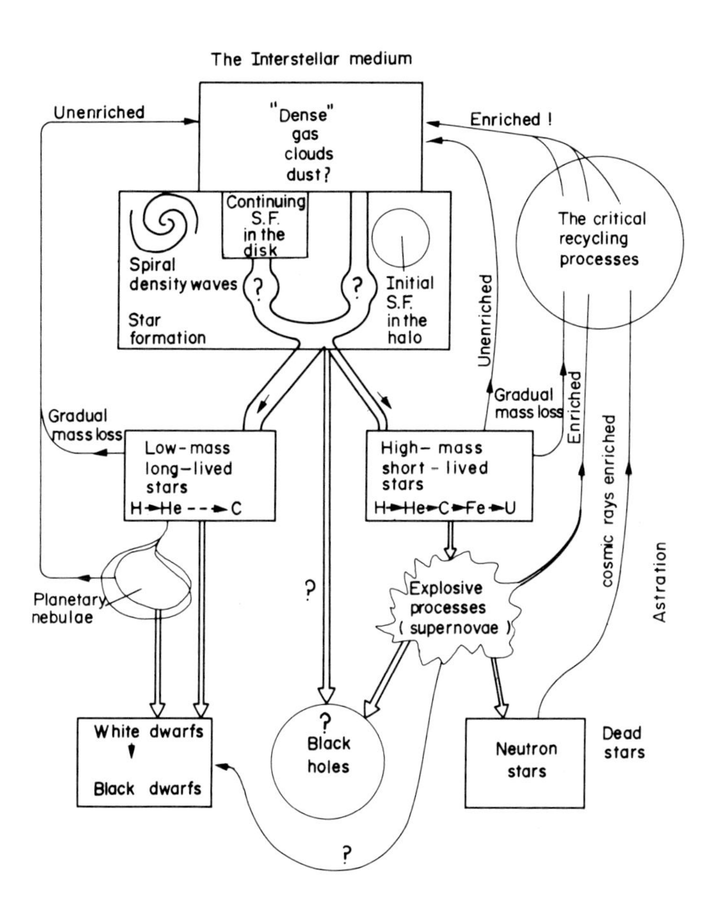

Figure 4-8. Astration and Cosmic Recycling
Note that the recycling of enriched material depends on contributions from high-mass stars only. Note also that in the halo of spiral galaxies like our own, and in elliptical galaxies, star formation is generally cut off quite early, so that enrichment of the interstellar medium by astrated material does not lead anywhere there.

An extreme case has been envisioned in the science-fiction novel *The Black Cloud*, by the noted astrophysicist Sir Fred Hoyle. The complexity of the possible interactions between a dense interstellar cloud and the solar system, is described in some detail therein, as well as Sir Fred's views on the competence of politicians facing a global crisis. Although the book was written over 20 years ago, the scientific aspect has held up very well: very complex molecules have indeed been shown to exist in interstellar clouds. Surprisingly perhaps, the technological predictions were not so good: both electronic computers and air transportation advanced more rapidly than Hoyle foresaw.

This might conclude our review of the links between galactic and solar system history. What, however, of the future? Once it was recognized, following Copernicus, that the stars were very distant, and might well be suns, it was speculated that planets might accompany them, and that intelligent life might exist on some of these. At the time of Giordano Bruno, such speculation was not well received.[5] In recent years, these speculations have become popular, and attempts have been made to give them a scientific form. Both the likelihood of and the optimum conditions for communication with intelligent civilizations elsewhere in the Galaxy have been estimated in various ways. Some searches have been made at various radio wavelengths, unsuccessfully. Despite the gross inefficiency of material transportation over interstellar distances, the Pioneer and Voyager spacecraft, destined to leave the solar system, carry messages describing our own civilization, in case they should meet anyone. Of course, many of our radio and television signals, racing ahead at the speed of light, carry other descriptions, perhaps not so flattering, of our culture. Will there be a reply? That will take centuries, perhaps millenia, unless another intelligent civilization exists very close by. Will we receive a message during the lifetime of our own apparently unstable technical civilization, and could such a message affect its lifetime? What indeed would be the consequences of contact with an extraterrestrial civilization, probably much more advanced than ours? Many distinguished people have discussed these questions, but really we do not know. In the solar system, life of any sort has only been found on Earth, so the statistics on which to judge theories of the origin of life are inexistent. Interstellar travel seems very remote indeed. Perhaps the best resource we have is to wait and "listen".

5

The Stability of the Terrestrial Environment

> Que me conseillez-vous d'aller visiter? demande-t-il.
> La planète Terre, lui répondit le géographe.
> Elle a une bonne réputation . . .
>
> Antoine de Saint-Exupéry[1]
> *Le Petit Prince*

WE have at this point completed our review of the very largest scale elements which constitute the environment of the planet and which determine — in various ways — our environment on Earth. We have seen how the universe as a whole evolves, starting from the violence of the big bang. We have seen too how the Galaxy evolves, its own evolution the result of the sometimes violent evolution of billions of stars. Constancy does not exist, and yet, the fact that we are here proves that the terrestrial environment has at least remained appropriate to the continuing evolution of life, over most of the last few billion years. In the chapters which follow this one, we shall examine the nearby elements of the terrestrial environment — the Sun, the Moon, the other planets of the solar system. We shall see that it is probably in these nearby bodies, and on our planet itself, that the causes of the principal fluctuations in our environment on Earth are to be found. Before going ahead to these questions, however, it seems useful to review briefly what we know of the history of the Earth environment, and to develop a few basic concepts which we shall need in the rest of this book.

Let me begin by defining the most immediately important part of our environment, what I would call the domain of action of Man. I think of it as an extension of Man's "natural habitat". This "natural" habitat includes all places where humans have been able to live and reproduce over at least a few generations. It is essentially limited to land (and lake) surfaces, from sea level (or slightly below) to quite high altitudes, perhaps as high as 7 000 meters. It by no means includes all land surfaces, nor does it include the ocean depths, which nevertheless harbor life. The domain of action of Man goes much further: it includes the oceans, from the surface to the bottom, all land areas, all the atmosphere, and a few kilometers down into the crust. In the last 20 years we have seen the extension of the human domain of action into space, to the Moon, and even out to the giant planets. We have not penetrated much deeper into the interior of the Earth, however.

A full description of the domain of action of Man would include geography, the positions of the continents, the volume of the oceans, the extent of the ice caps, the strength and direction of the magnetic field, the composition of the atmosphere, the oceanic and atmospheric circulation patterns, the "climate", the "ecology"; also the distance of the Moon and the precise parameters of the Earth's orbit around the Sun. All of these aspects of the domain of action of Man — what today constitutes our environment — have changed during the more than 4 billion years of the Earth's history, and some of them have changed significantly, even quite recently, since the emergence of Man some few million years ago. All of these aspects of our environment are inter-related, within what has been called the "geosystem".

Even today our description of this geosystem is far from complete. For the past we obviously have to work with ambiguous and incomplete evidence. This is often true even for the relatively recent past and for the present. Thus, while meteorological data have been compiled in numerous locations for many decades and even centuries, a global coverage was not feasible before the advent of the space age and the launching of meteorological satellites. The first of these were necessarily experimental, and it is only now, with the launching of the Global Observation System (which includes the European Space Agency's Meteosat as well as American, Soviet and Japanese satellites) that permanent continuous well-calibrated observations are becoming available. Nevertheless, the data on the oceans remain extremely spotty, and while ambitious international programs are being discussed, political problems, which have emerged in the Second Law of the Sea Conference,[2] severely threaten to hamper oceanographic research in the critical regions above the continental shelves, so that the situation may not improve in the near future. Even when we do have a continuous series of measurements going back several decades — measurements, say, of insolation or temperature in a particular place — doubts about the comparability of old and new data may arise, as observers and instruments are changed. Even without attempting to establish a history of the past, the compilation of such series of measurements is essential to a determination of the "present" state of things, since short-term fluctuations over periods of a few years are common. Thus it has been pointed out that the recent severe drought in the Sahel region of Africa is by no means an unprecedented phenomenon, and cannot be said to herald a permanent change in climate.

When we do try to establish past history, we work necessarily with incomplete data, and often must rely on indirect evidence concerning the state of things. Quantitative meteorological observations using instruments began only about 300 years ago, in Europe, but written historical records bearing on climate and other aspects of the environment go much further back, as much as 5000 years for China. In some cases actual rainfall measurements are available. In general, exceptional occurrences, such as earthquakes, volcanic eruptions, unusual warm or cold periods, the freezing of rivers, eclipses of the Sun or Moon, destruction of villages by advancing glaciers, prolonged droughts, . . . are reported and described in some detail. Routine records of agricultural activity, including for example the dates of wine harvests in France or the blooming of cherry trees in Japan, provide valuable information to the historian of climate. So do records of the shorelines of lakes, the limits of glaciers, the prevalence of sea ice. Historical data of this sort can be combined with other types of information, such as can be obtained from tree rings or pollen deposits, to establish how climate has varied in different parts of the globe over the past few thousand years. Correlations between these variations seem to exist in many areas, indicating global changes in climate. Obviously such changes have affected the course of human history, and various authors have elaborated on this theme. One must examine this literature with care. There may be good arguments to show for example how the migrations of peoples have been driven by changes in climate, and once a "model" for this is devised from independent evidence concerning the two, one may well be tempted to infer climatic history from the history of migrations. Such inferences, however, are only as good as the model. The danger of circular reasoning is clear, and it has not always been avoided.

Leaving now the effects of global climate changes in historical times, what of the causes? We shall discuss this in some detail in the rest of this book. Obviously the historical period of 8000 years at best is but a moment compared to the age of the Earth. We shall be trying to understand the changes that have taken place over both long and short time scales compared to this whole period. For historical times we of course have more complete data. Thus, solar activity, as manifested in particular by the sunspot cycle, has been suggested as a significant influence on climate. Since Galileo, i.e. since over 300 years, records of sunspots have been kept. Apart from isolated Chinese observations such records are not available earlier. Recent studies of the historical record by Colorado astrophysicist Jack Eddy suggest that the sunspot cycle has itself changed during the last few centuries, and that these changes may be correlated with changes in climate. We shall come back to this when we discuss the Sun.[3]

Now let us briefly summarize the history of climate over the past 5000 years, as established from the many types of data involving very different disciplines of research. In general the fluctuations in this period have been small and the environment has been fairly stable. The fluctuations that have occurred are by no means negligible however; much smaller fluctuations over the past few years have been sufficient to generate considerable worry in politico-economic-agricultural spheres.[4] This is to be expected, since in most regions of the world, present food production is at best marginally adequate to prevent starvation. The difference between famine and survival at times depends on the surplus production of what could be called the New World "Anglo-Saxon" countries — North America, Australia (although many or most of the farmers are of German, Scandinavian or Slavic origin). Even "small" climate fluctuations in these regions, in the USSR, or in South or East Asia, can be a matter of life or death to millions of people.

It is not yet clear whether the climate or weather fluctuations of the last three decades point to a new trend, although it does appear that a century-long trend of warming, with receding glaciers from 1850 to 1940 has been halted and perhaps reversed. Other aspects of the environment have changed in the past century. There has been a clear increase in the carbon dioxide content of the atmosphere, no doubt the consequence of the burning of fossil fuels, perhaps also due to extensive deforestation. In recent decades the concentrations in the atmosphere and oceans of many man-made substances have also increased, and there has been considerable concern about the consequences of such pollution. Examples include DDT and its effects on living organisms, or the chlorofluoromethanes (freons) and the destruction of ozone in the upper atmosphere. We shall say more on these questions later. Pollution by agricultural and industrial practices includes the dispersion of various types of particles (aerosols) in the atmosphere, but whether this is significant on a global scale is not certain. There are many natural sources of aerosols too, and some of these, like volcanic eruptions, are variable. Establishing the relative impact of human activities is by no means easy; in principle, as we shall see later, the release of aerosols on the one hand and that of carbon dioxide on the other, should have effects on global climate.

To summarize, there are small fluctuations in climate today, which may or may not signal a significant change in the trend of the past hundred years. The present period is one in which anthropogenic (i.e. man-caused) changes in the environment (pollution, etc. . . .) are beginning to be comparable with natural fluctuations such as those associated with volcanic or solar activity. Certainly if anything like the present rate of growth of technological activity is maintained over the next few decades, without stringent "anti pollution" measures, substantial impact on the environment — including climate — is to be expected. This will be discussed further in the last chapter.

One point not always recognized is that even the fluctuating climate of today is warmer — at least at North American and European latitudes — than it has been at many times in the historical past. Average temperatures in the 1940's were 0.5°C higher than in the 1880's. Was the warming trend of the late 1800's linked to the abatement of volcanic activity during that period? Or was it linked to the reinforcement of the solar activity cycle, as argued by Eddy? Both phenomena, and many others, may be involved. Looking further back in history, we note in particular the "Little Ice Age" often given as covering the entire period from 1430 to 1850, and culminating between 1610 and 1640. During this period, average temperatures were as much as 1°C lower than in the 1940's, rivers and lakes froze much more commonly in Western Europe than they do now (witness the Breughel paintings), and glaciers advanced, at times quite rapidly, destroying Alpine villages. Were this sort of climate to re-establish itself rapidly, there would be striking changes in some living patterns. However, we cannot assume that this would necessarily lead to a world food crisis, as production in the major world "bread baskets" is probably limited more by the water supply than by the temperature. There is some evidence that production in North America would actually increase with a fall in temperature.

Proceeding back in history, prior to the Little Ice Age, we note the so-called Little Optimum period going roughly from A.D. 800 to A.D. 1200, during which significantly milder conditions prevailed. Global average temperatures may have been 1°C higher than in the 1940's, and at high latitudes the difference may have been as high as 4°C. This was the period of the Viking explorations and settlements in Iceland, Greenland, and along the northern coast of North America. Generally, there seems to be agreement between the historical data for Europe, and tree-ring data for North America, regarding the succession of climates from the Little Optimum to the Little Ice Age. The evidence from China, on the other hand, seems to indicate that both of these periods occurred earlier there than they did in the "West". Much work remains to be done on these questions.

Going still further back, we again encounter a cooler period between 900 and 400 B.C. (3000–2400 B.P. before the present), in Europe and North America, possibly also in China at least during the beginning of this period. The earliest historical records, combined with other evidence, also reveal a rather warm period, often known as the Postglacial Optimum, about 6000 to 5000 years ago, when temperatures were a few degrees higher than at present. Of course, it should be noted that where we have characterized periods as "warm" or "cold", this is an extreme simplification of a complex reality involving precipitation, humidity, evaporation, and wind patterns, as well as temperature. Moreover, except for the last few hundred years for which a record of temperatures as such is available, the indicators of climate used are seldom determined solely by temperatures, but usually rather by the combined influence of many different factors. Finally, it should be emphasized that all these variations of the last 6 000 years, large as they may be compared to the fluctuations of the past few decades, remain quite small compared to some of the earlier variations.

Before we consider the much longer prehistorical record, let us note that other elements of our environment, apart from climate, have changed measurably in historical times. Observational astronomy has been a well-established science since at least 2000 years, and even older eclipse records are usable. These records show that the rotation of the Earth has been gradually slowing down, so that the length of the day is increasing by about 2 milliseconds per century. One can show that the Moon's orbit must at the same time be increasing in size, and the month getting longer. This phenomenon is understandable in terms of the friction exerted by the tides on the rotating Earth, as we shall see in a later chapter. It is confirmed by examination of the much longer fossil record. Certain corals which existed in the Middle Devonian, about 370 million years ago, have daily growth rings which also show annual variations related to the seasons; there seem to have been about 400 days in a year at that epoch, with the day lasting less that 22 (present) hours.

Volcanic activity is of course a sporadic affair, which may well have an impact on global climate. This is only one aspect of the internal activity of the "solid" Earth. Both volcanic and seismic activity are related to the processes which must also account for the spreading of the sea floor and the drift of the continents. These processes were discovered from the prehistoric evidence rather than from the historical record, but they are directly observable now. Moreover, definite links exist between earthquakes and what is known as the "Chandler wobble" in the rotation of the Earth. We shall come back to this in a later chapter; the point here

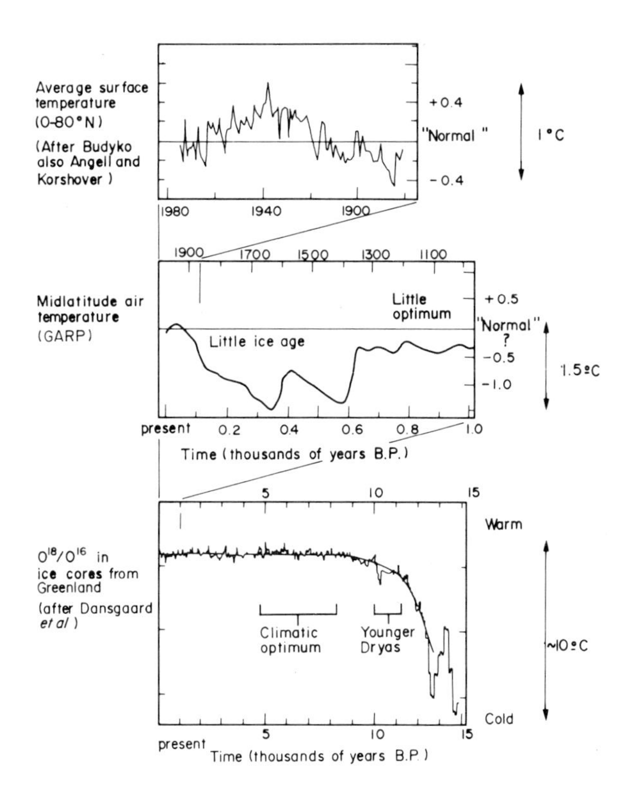

Figure 5-1. Recent climatic history, i.e. since the end of the last glaciation.[5]

is that since the Earth is not a perfectly rigid body, the maintenance of such a wobble requires an energy source. Major earthquakes do indeed correspond to measurable if very small displacements of the Earth's axis of rotation, and earthquakes can help to keep the wobble going, just as the wobble can sometimes trigger earthquakes.[6] The energies involved (about 3 10^{11} kilowatt-hours in an average year) are small compared to the heat flow from the interior of the Earth (more than 500 times larger), not to mention the energy flow from the Sun (nearly 1000 times larger still). However, since the release of such energy is concentrated in time and place, it can have a devastating effect on human life.

Still another fluctuating aspect of the environment is the Earth's magnetic field. This must be the result of motions in the molten part of the iron-nickel core of the Earth, functioning as a dynamo, but it is not yet fully understood. These motions must interact with the Earth's rotation (which as noted earlier is gradually slowing down), with the Moon's orbit, and with motions in the Earth's mantle, so that the situation is extremely complex. Over the past decades during which measurements have been made, magnetic field patterns on the Earth's surface have drifted quite rapidly. When the prehistoric record is examined, using fossil magnetism (which we shall describe shortly), it is found that the magnetic field of the Earth has in fact reversed itself many times. This certainly must have had important consequences for the structure of the upper atmosphere and the magnetosphere, through rather well understood physical mechanisms which we shall study in the next chapter. Some researchers also suggest links between the magnetic field and weather and climate.

Now let us look at the history of the planet in the longer time scale. Here historical records and tree rings are no longer of any use. We depend on the geological record, with dating by isotope ratios, to take us very far back, and even then the record becomes extremely spotty when we go back more than a few hundred million years. It would be risky to attempt to describe the climate of the Precambrian — the first 4 eons (billion years) of the Earth's history. The important point is that there is evidence for the existence of life quite early in this era, some 3.5 eons ago, and the continued existence of life since then implies a certain degree of stability in the environment; there must have been liquid water. During much of this period, however, the atmosphere may have contained only very little free oxygen, and so the surface of the Earth may not have been shielded from the solar near ultraviolet radiation, which at present is absorbed by the ozone layer. When we reach the Palaeozoic Era, starting only 600 Myr (million years) ago, the geological record becomes much richer. Well preserved fossils are found in much greater variety, life evolves rapidly from relatively simple forms to the fantastic diversity found today, not only in the oceans, but also on land and in the air. This too is the period in which free oxygen appears in the atmosphere, no doubt the product of photosynthesis by algae. Apart from this spectacular explosion of life forms, two grandiose phenomena profoundly modify the terrestrial environment over the last 600 Myr. They are the continental drift, and the coming and passing of the great ice ages.

Considerable evidence for continental drift was marshalled by the German meteorologist and explorer Alfred Wegener over 60 years ago,

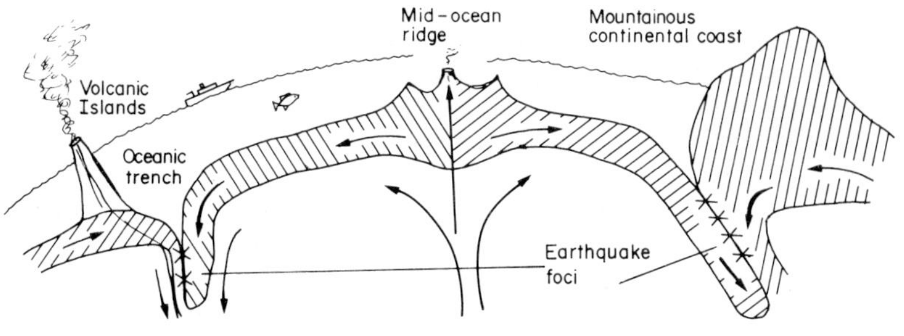

(a) The crust and the top 700 km of the mantle
(vertical scale exaggerated)

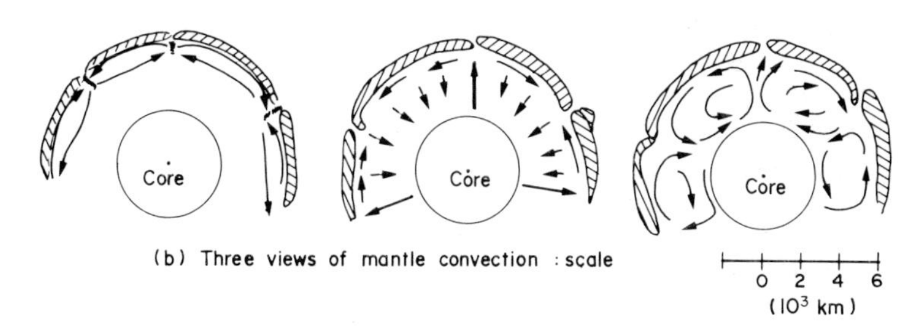

Figure 5-2. Convection in the Earth's mantle, sea-floor spreading and continental drift. Three views of mantle convection are shown.

Left: Shallow convection, not extending more than 700 km into the mantle.

Center: Rising material is concentrated in compact thermal plumes, producing "hot-spots" at the surface. The falling material is spread out in the rest of the mantle.

Right: Deep convection extends all the way through the mantle to the core.

but only very recently, following what has been called "a revolution in the Earth sciences", has this concept been generally accepted. The evidence includes not only the close fit between the West African and South American coastlines, but also an abundance of correlations between geological formations, and the various species of reptiles and mammals, that exist in regions now separated by thousands of kilometers of ocean. It does indeed appear that about 300 Myr ago, all of the present continents were grouped together in a single land mass, often given the name Pangaea, surrounded by a universal ocean. This then broke up into two groups, the northern group "Laurasia" including North America and most of Eurasia, the Southern group, Gondwanaland, being made up of South America, Africa, the Indian subcontinent, Antarctica, and Australia. In the late Palaeozoic, some 300 to 250 Myr B.P., Gondwanaland must have been located around the South Pole, and was the site of widespread glaciation lasting some tens of millions of years. The geological traces of this glaciation were strong evidence for Wegener that continents like Africa had not always been located far from the pole, and that continents could indeed drift. Further breakups occurred: South America, Africa, India and Antarctica separated first, India colliding subsequently with the Eurasian landmass, producing the Himalayas, and North America also separated from Europe. All of these changes radically modified the distribution of land and sea with latitude, thus certainly changing local climates (which helps to explain the existence of coal, laid down by tropical forests, in both Spitzbergen and Antarctica), and also affecting global climate.

The mechanism accounting for these titanic changes is the spreading of the sea floor, starting from the mid-ocean ridges. "New" molten rock wells up from the Earth's mantle along a colossal continuous chain of mountains, which was discovered by oceanographers only some 25 years ago. The rock then spreads out, pushing apart the continents which float like rafts on what are called the continental plates. These upwelling lava flows have been photographed. As the lava emerges it cools rapidly in the ocean water, and it becomes magnetized, the strength and direction of the prevailing magnetic field of the Earth thus being recorded in the ocean floor. The fresh lava coming out of the mid-ocean ridge keeps pushing away this cooled rock, so that along the ocean floor we can find a continuous record of the terrestrial magnetic field. The records from the various oceans show that the sea floor has spread out at a different rate in each of them, but that a common sequence of close to 200 reversals of the magnetic field over the past hundred million years can be established.

What is the energy source for all this? At the Earth's surface, we observe an average heat flow of about 60 milliwatts per square meter from the interior. Nearly half of this heat flow is due to the continuing radioactive decay of potassium-40 and various isotopes of thorium and uranium, left-overs from the supernova explosion(s) which must have preceded the birth of the solar system. However, these elements and the associated heat source are concentrated in the Earth's crust, *above* the mantle, while convection in the mantle requires heating from *below*. Heat is indeed available from below: the central core, both in its solid inner part and in its molten outer part, is at a temperature between 4000 and 5000°K.

The question may be raised as to how it got to be so hot. Most geophysicists believe the core to be largely made up of iron, nickel and cobalt, with some silicon or sulphur mixed in; none of these is a radioactive heat source. Potassium-40 is, and there is controversy as to how much of it can be present in the core, heating it. At any rate, the core is surely quite different in composition from both the mantle and the crust, and it seems clear that this together with the high temperature depends on the processes by which the Earth was formed. Were the elements like iron and nickel the first to condense out of the planet-forming cloud, so that the mantle collected around an already-formed core? If so, some of the heat now in the core may be a residue of the heat produced when it was formed, heat produced by the release of gravitational energy as more and more metallic grains and later meteorites fell onto the growing condensation. Alternatively, did the Earth form with its minerals initially distributed more or less uniformly throughout its volume? Many variations of this view have been proposed, one of the earliest by the American geophysicist and geochemist Harold Urey over 20 years ago. According to this view, after a relatively cool beginning, the outer layers of the young Earth were heated by meteorite impacts, by radioactivity, and perhaps by a strong solar wind, with the result that elements like iron melted. Being dense, these elements sank through the mantle towards the Earth's center, with possibly a rapid collapse forming the core. This would have released substantial amounts of gravitational energy, and much of the heat produced may still be in the core today. Moreover, the process may still be continuing, and the core still growing.

Clearly, we still have a lot to learn about the inside of the Earth. There are a number of conflicting theories, the evidence is often incomplete or ambiguous, and the differences are not simply matters of detail. Whichever view turns out to be correct, we can say that the prenatal history and the birth pangs of the Earth are still with us. Heat and magma pour out at the mid-ocean ridges: in Iceland, this geothermal energy is tapped to heat homes and generate electricity. Where the spreading sea floor is forced under the continental plates, the Earth trembles and volcanoes erupt. The solid Earth is a young and active Earth.

Now let us look at the great ice ages. Over the last 600 million years, since the Cambrian Era, widespread glaciation has come and gone. The late Palaeozoic glaciation, 300–250 Myr B.P., has already been mentioned. We believe that it may have lasted from 30 to 50 Myr, but we have little information on the details of how it varied during that period. We recall that at the time, Gondwanaland was at the South Pole, and this polar positioning of a major continental mass may be necessary, although not

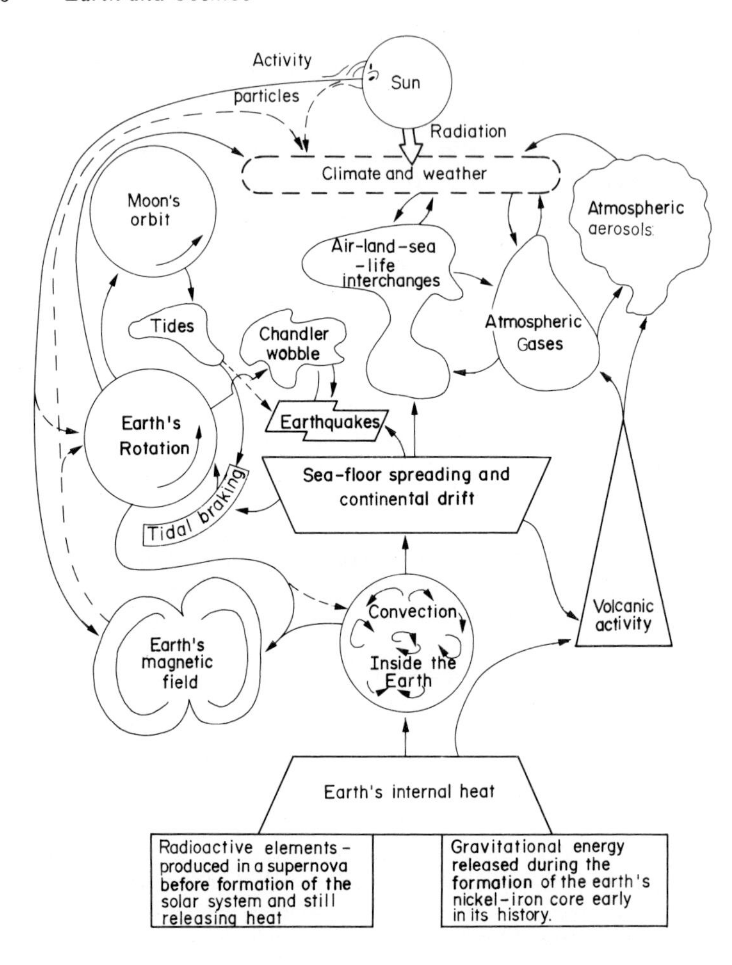

Figure 5-3. Internal Earth processes and their relations with other environmental processes.

sufficient, for the development of major glaciation. The ice cap of this glaciation must have been enormous, much larger than present ice caps in Greenland and Antarctica, probably larger too than the ice cap that covered much of North America and Europe about 18 000 years ago. Nevertheless, over 90% of the time during the last 600 Myr, the Earth has been completely ice free, with global average temperatures several degrees higher than now, and with a substantially milder, even what today we would call subtropical, climate at high latitudes. The fact is that we are now living in an ice age, which began only 2 or 3 Myr ago.

Some data on climate is available for the last 60 Myr, much more for the last million years. One of the most valuable methods, pioneered by Cesare Emiliani of the University of Miami involves measuring the oxygen isotope ratio O^{18}/O^{16} in the carbonates of fossil seashells (of organisms of the foraminifera family), found in sediments of the ocean floor. Because of the difference in mass between the two isotopes, otherwise chemically identical, the relative amount of O^{18} locked up in carbonates depends on the temperature at the time the carbonate is formed. Thus we have a thermometer of ancient times, which we can read by the analysis of ancient sediments. Deep sea drilling from the oceanographic research vessel Glomar Challenger has yielded cores of sediments covering the last 70 Myr B.P., and these reveal a gradually decreasing temperature. It has been suggested that this corresponds to the Antarctic continent gradually nearing its present position at the South Pole, with a steep drop in ocean temperatures once sea ice began to form there, but other factors may also be involved.

The formation of large volumes of ice complicates the oxygen isotope method as applied to sediments, since the overall isotopic composition of ocean water then changes. This however creates new possibilities, because the isotope ratio can be measured in ice cores obtained in Greenland or Antarctica, giving information on the climate in these regions over the past. An ice core 1387 meters deep obtained at Camp Century in Greenland shows the fluctuations in both interglacial and glacial periods of the past 120 000 years, and deeper cores may yield records up to 1 or 2 Myr. As ice caps grow or recede over continental areas, the amount of water left in the oceans changes, and shorelines advance or recede in response to the fall or rise of the ocean level. For example, one might expect many important cities to be flooded were the Antarctic and Greenland ice caps to disappear. Actually, the situation is not so simple. Under the weight of a fully developed ice cap, a continent sinks a bit into the mantle of the Earth. When an ice cap, such as the one which recently covered much of northern America, melts, there is more water in the oceans, but at the same time the land previously covered by ice begins to rise. Thus northern Europe is still recovering from the effects of the last glaciation, and it may manage to stay above water even when all the ice in Antarctica melts.

At any rate, thanks to geology, thanks to the isotopic analysis of fossil foraminifera and of ice cores, thanks also to the study of fossil pollens, we have a fairly detailed history of the current ice age, which began between 3 and 2 Myr ago. There is a suggestion, from carbon isotopic data, of a corresponding major change in the ecology, with extinction of many plant species and a significant drop in the biomass. This current ice age is in fact made up of a succession, at roughly 100 000 year intervals, of cold glacial periods and moderately warm interglacials, during which however the ice caps of Greenland and Antarctica have never disappeared. We are presently living in an interglacial, the last glacial maximum having occurred about 18 000 years B.P., the last interglacial optimum, already mentioned, about 6000 B.P. The volume of ice may vary by a factor of 3 between glacial and interglacial maxima, leading to changes in sea level of about 100 meters. At some epochs glaciers appear to have begun their advance extremely suddenly, and volcanic activity has been suggested as the triggering agent. In general however the growth of glaciation seems to have been more gradual than recession. We should note that many "short-term" (i.e. 100—20 000 year) fluctuations are found, some of them quasi-periodic. Recently, the Milankovitch theory, which links some of these fluctuations and the 100 000 year rhythm to small semi-regular but fully predictable variations in the astronomical parameters of the Earth, has been revived. We shall discuss this in a later chapter.

We have seen that the stability of the world environment, and in particular of climate, is a very relative thing. Since much of the rest of this book will be concerned with the elements of climate, let me try to define this concept more precisely. Climate is some sort of average over weather. The weather is an essentially local phenomenon, the state of the atmosphere in a given time and place. We can average the weather in different ways. For example, we can obtain elements of the local climate (possibly a micro-climate if it differs substantially from other local climates found nearby), by performing time averages of the weather at a particular place on Earth. One might be tempted to define the local climate in terms of the annual average temperature and rainfall, for example, and such figures can be useful. However, because the Earth has seasons, because the distribution of land and sea areas, which have very different properties with regard to heat and humidity, is irregular, so simple a characterization of climate would be highly misleading. Let us consider the shores of the Atlantic. The annual average temperature is not very different in Boston from what it is on Valencia Island off the West coast of Ireland, but one could hardly say that they have the same climate, extreme weather being far more commonly encountered in Boston. Even considering more restricted time averages, say performing month-by-month averages (possibly

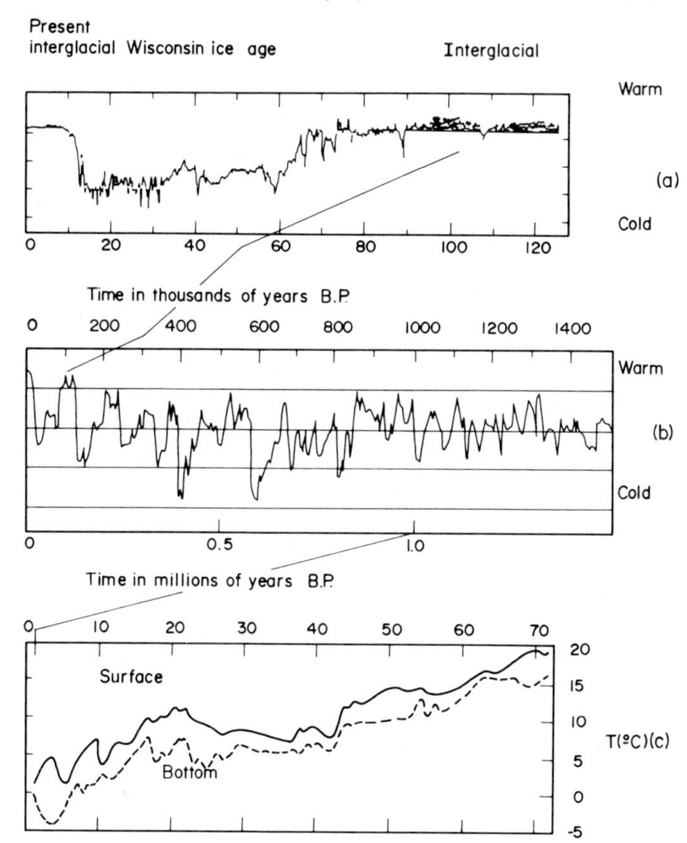

Figure 5-4. Climatic history, deduced from the oxygen isotope ratio.[7]

a) The last 130 000 years, as recorded in a 1387-meter ice core from Camp Century, Greenland. (after Dansgaard et al.)
b) The last 1.5 million years, as recorded in the fossils of foraminifera, in a sediment core from the floor of the Pacific Ocean (after Shackleton and Opdyke.) Note the alternation of glacial and interglacial stages.
c) The last 70 million years, leading up to the current Pleistocene ice age. Again foraminifera microfossils in deep-sea sediment cores have been used. Bottom- and surface-living species can be distinguished. (after Shackleton et al.)

combining say April measurements from several years to increase the sample), will not yield a full determination of climate. The point is that the climate of Boston in April is one that includes weather types ranging from snowfalls of 1-meter and subfreezing temperatures, to hot days with 33°C temperatures. While the average April temperatures and precipitations are about the same on Valencia Island as in Boston, these extremes practically never occur there. A full characterization of a local climate not only has to include both average values and variances of say temperature, precipitation, wind velocity, cloudiness and cloud type, etc. . . . combining data from several years taken month-by-month, but it also must include information on the frequency and duration of particular weather types.

In contrast to the local approach, we can try to see how the *global* climate could be characterized. We could of course imagine computing an average temperature, humidity, precipitation, etc. . . . for the whole Earth (although not all the data is really available). I believe it will be more instructive and meaningful to begin by considering only temperature, and to look on the Earth from the vantage point of a distant astronomer. Then we can perform no measurements directly on the Earth itself; we can only consider radiation. We note that the Earth reflects a fraction of the visible and near infrared solar radiation that it intercepts. This fraction, which is called the global albedo, is about 30% for the Earth, intermediate between the low values of about 6% found for Mercury and the Moon, and the high values of about 75% found for Venus and Jupiter. At the Earth's distance from the Sun, the solar flux (above the atmosphere) is about 1360 watts per square meter (this is called the solar constant). Since the Earth has a cross-section of about $1.27\ 10^{14}$ square meters, and absorbs 70% = (1-albedo) of the intercepted flux, the Earth must be continually absorbing about $1.2\ 10^{17}$ watts of solar power. Obviously the Earth is being heated by this, and since conditions on Earth remain roughly constant, there must be equivalent cooling processes.

The only practical cooling process for the whole Earth, isolated as it is in interplanetary space, is electromagnetic radiation. If our hypothetical distant astronomer observes the Earth with instruments sensitive at longer wavelengths, in the middle and far infrared (wavelengths from say 5 up to 50 micrometers) he (or she!) will find that $1.2\ 10^{17}$ watts are continually being radiated by the Earth, at these longer wavelengths. Actually, although some such observations have been made from artificial satellites, coverage has not been continuous, and much remains to be done to improve their accuracy. Still it is reasonably well established that the amount radiated away in the infrared is practically equal to the amount absorbed in the "visible", within the accuracy of the measurements. As mentioned earlier, the rate of energy outflow from inside the Earth (geothermal power), like the rate of energy production by Man from fossil or nuclear fuels (i.e. not from present solar input, which is the source of hydroelectric power), is much much smaller. Certainly a distant astronomer would have no evidence of the existence of such processes, from the radiation measurements we have mentioned so far. Similar measurements have been performed for other planets, and generally this balance between the input of solar "short-wave" radiation, and the outflow of planetary "long-wave" radiation, seems to hold. Jupiter is an exception: the outflow seems to be roughly twice as high as the solar input, and for this situation to persist, there must be release of energy inside the planet, presumably gravitational energy from slow continued contraction of the planet.

This leads us to compute a first parameter of global climate in the following way. We assume that the long-wave radiation is thermal radiation, i.e. that the surface (if any) and atmosphere (if any) of the planet is radiating simply because it is hot, or rather because it is not at absolute zero. Lacking any other information at this point, we assume further that the planet radiates like a "black body" or perfect radiator, i.e. it radiates in a way that depends only on the temperature and not at all on the other properties of the surface or atmosphere. With these assumptions the flux of long-wave radiation from the planetary surface is given by the Stefan-Boltzmann law: $F = \sigma T_{eff}^4$, where σ is the Stefan-Boltzmann constant, and T_{eff} is the effective temperature. Note that this calculation is simplified by the fact that the thermal long-wave domain and the solar short-wave domain of the spectrum are well separated, so that we can consider them separately, and this is simply due to the fact that planets are much cooler than the Sun. We assume now that the thermal flux F is emitted by the entire surface of the Earth, not taking into account real temperature differences related to the fact that the solar energy input is unevenly distributed. We are after all looking for a global average. The result is that the equation:

Solar input = long-wave thermal outflow

becomes: Solar Flux × (1-Earth Albedo) × Earth cross-section
= Earth surface area × Black body flux

or

$$F_{\odot} \times (1-A) \times \pi R^2 = 4\pi R^2 \sigma T_{eff}^4$$

and the actual value of the Earth's radius is irrelevant. (Thus the distant astronomer need not measure it). Our distant astronomer then finds that:

$$T_{eff}^{4} = (1\text{-}A)F_{\odot}/4\sigma$$

and putting in the numbers, it turns out that $T_{eff} = 255^{\circ}K = -18^{\circ}C$ for the Earth. This may appear surprisingly low. Indeed, the global average temperature for the Earth's surface is about 288°K, but the distant observer cannot know this from the data thus far examined. Clearly restricting ourselves to a single parameter for the definition of global climate can be extremely misleading.

What more can our distant astronomer do? One possibility is to study the spectrum i.e. the distribution of intensity with wavelength, of the

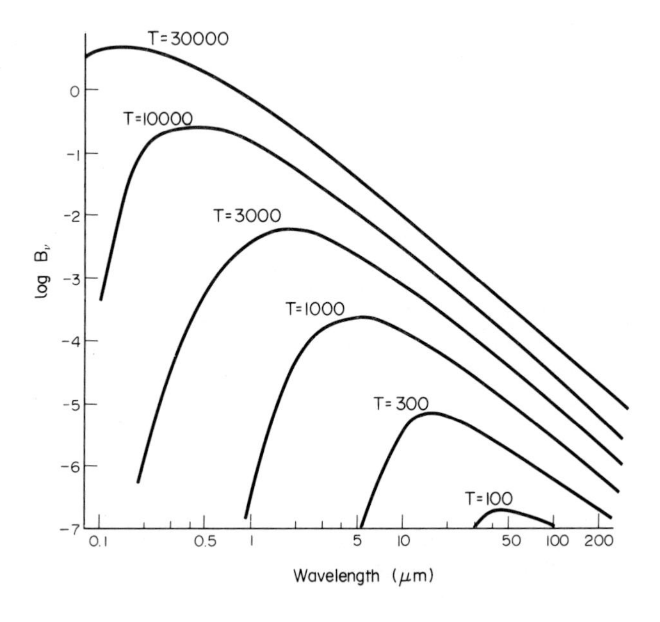

Figure 5-5. The spectrum of black body radiation, for various temperatures. The ordinate is the logarithm of the Planck function, which is given here in terms of the energy flux per unit frequency interval, in Watts per square meter per Hertz.

planetary thermal radiation. Detailed examination of the long-wave spectrum reveals that the planet does not radiate exactly like a black body; the overall form of the spectrum is somewhat different, and there are lines and bands due to various molecules, atoms, and ions. Particularly prominent are H_2O and CO_2, but many others play important if apparently minor roles. These lines and bands are for the most part in absorption, (as seen by our distant astronomer; looking up from the ground, we see them in emission against the dark sky). This suggests that the thermal radiation from the planet comes in fact from a gaseous atmosphere in which the temperature decreases outward. At the wavelengths of the lines, the gas is more opaque, and so our astronomer "sees" the outer cooler layers and not the deeper warmer region. Thus, while he might not at this point deduce the existence of a solid surface, he would at least realize that the effective temperature is not the temperature everywhere in the atmosphere, and that in particular the temperature increases downwards. This however implies that energy is flowing upward through the planetary atmosphere; and yet the source of the energy is the Sun. The point is that since the atmosphere is relatively transparent to the "short-wave" radiation from the Sun, most of that radiation is absorbed at the surface itself. If there were no atmosphere (as is the case for the Moon and Mercury), that surface would on the average be at the effective temperature, and the corresponding long-wave thermal radiation would escape directly to space. However, the atmosphere is relatively opaque in the infrared; the long-wave radiation can traverse it only through a process of successive absorptions and re-emissions, and the re-emission of radiation by the molecules can be directed downward as well as upward. This tends to heat the lower layers of the atmosphere as well as the surface, until the difference between the surface temperature and the effective temperature is high enough to "drive" the energy flux through the partially opaque atmosphere. This heating effect is often called the "greenhouse effect". The glass panes in a greenhouse are indeed quite transparent to solar visible radiation, and not so transparent to the thermal infrared radiated by the plants, and this partial trapping of the infrared radiation does indeed contribute to keeping the greenhouse warm. However, the confining effect of the greenhouse enclosure, which prevents convecting air currents from cooling the interior, is probably at least as important, and so some researchers reject the term "greenhouse effect" and prefer to use the term "atmosphere effect" instead. There is no enclosure to prevent convective air motions from cooling the Earth's surface and lower atmosphere, and indeed the temperature rise is not as great as it would be if no motions were permitted. We might be tempted to call the Earth's

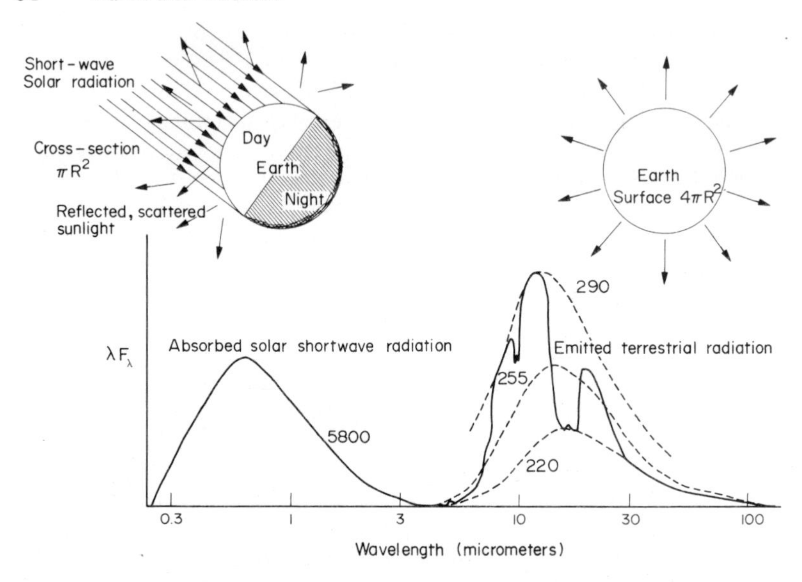

Figure 5-6. Global radiative equilibrium. The absorbed sunlight has roughly a blackbody spectrum corresponding to a temperature of 5800 K. The total emitted longwave radiation corresponds to the effective temperature of 255 K, but its spectrum is rather different from the blackbody spectrum because of absorption by atmospheric gases (water vapor and carbon dioxide especially). Over most of the spectrum only radiation from the cold upper atmosphere escapes, but we can "see" the warmer surface of the Earth through the infrared "window" at wavelengths 10 to 12 μm.

atmosphere a "drafty greenhouse". To complicate things still further, astrophysicists use the term "blanketing effect" or "back-warming" for what is essentially the same phenomenon, where absorption lines in the atmospheres of stars are concerned.

With the effective temperature, we had a single quantity characterizing climate, determined by 2 parameters — the solar flux, and the global albedo — possibly modified by a third — the rate of non-solar energy production on the planet, if it should become significant. This single quantity did not seem very useful. With the (drafty) greenhouse effect, we can extract a second quantity — the average surface temperature — which is related to the effective temperature in a way involving additional

Figure 5-7. The Earth (essentially Europe, Africa and the Atlantic Ocean) as seen from Meteosat, positioned at 36 000 km above the Equator on the Greenwich meridian, on 10 April 1978 at noon GMT.

a) Image in visible light, i.e. in reflected sunlight (left of Fig. 5-6). Note the brightness of clouds and of desert areas, contrasted with the rather dark oceans. Many surface features can be distinguished.

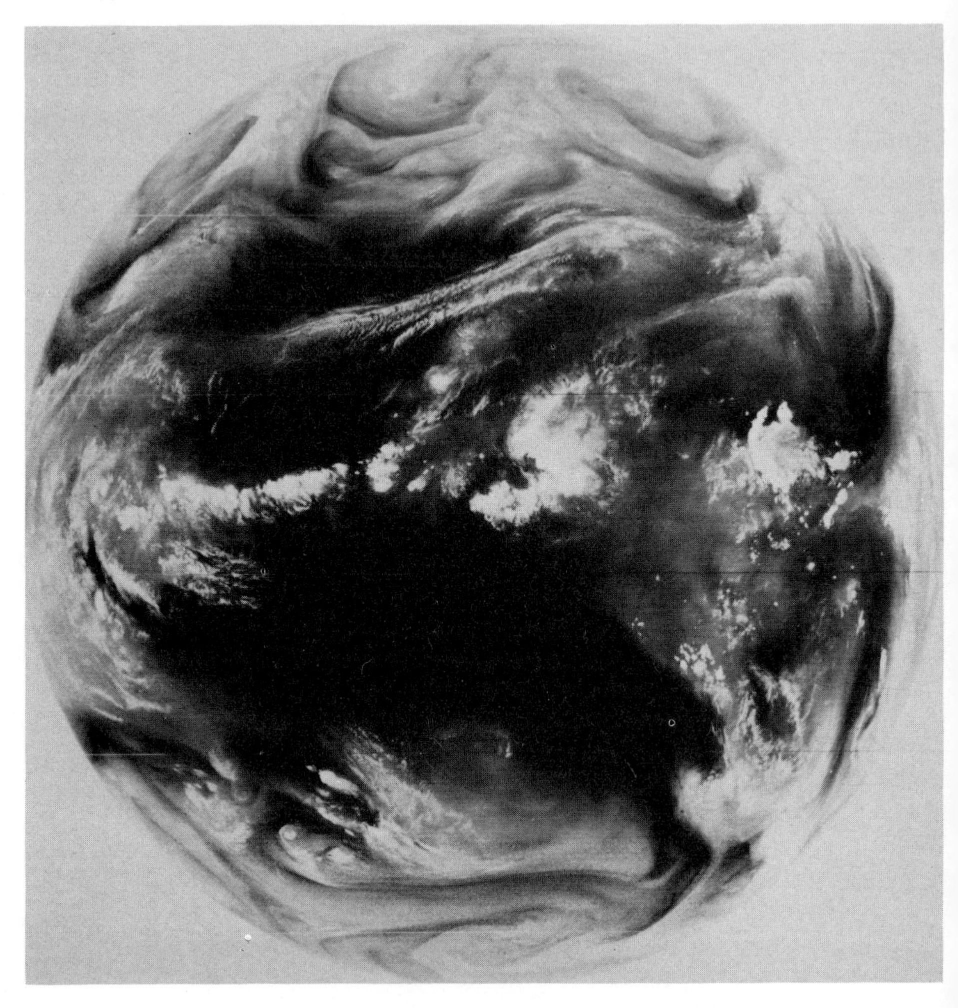

b) Wavelengths 5 to 7 micrometers, where water vapor absorbs radiation. The brighter (i.e. warmer) areas are printed blacker. Thus cloud tops appear whitest when they are highest in altitude and therefore very cold. Nowhere can the Earth's surface be seen, but where the image is darkest we are seeing down to about 5 km altitude where it is warmer. Water vapor is clearly distributed rather unevenly in the upper atmosphere.

c) The infrared "window" (10–12 micrometers), again with brighter warm areas printed darker, so that high clouds are whitest. Here the surface can be seen; note the land–sea contrast in temperature, and the different cloud heights. (Meteosat photos courtesy European Space Agency).

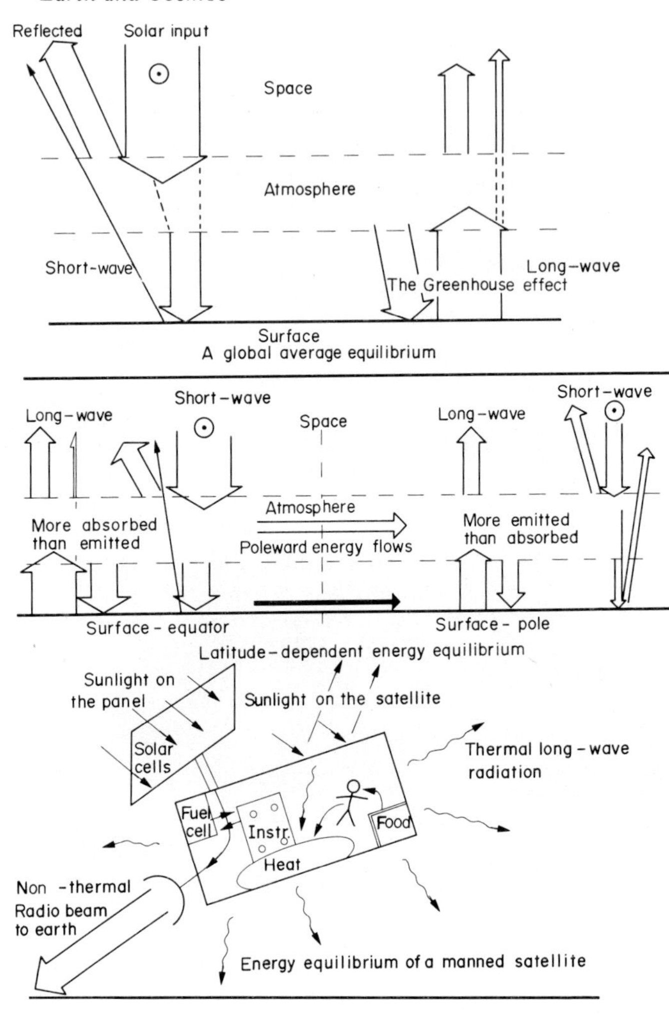

Figure 5-8. Different energy equilibrium situations.

parameters, one of which is the infrared opacity of the atmosphere. We are still dealing with global averages, but we have been forced to take into account the vertical structure of the Earth's atmosphere. We shall examine this in more detail in the next two chapters. Further on we shall see that we cannot fully understand global aspects of climate, without considering local aspects. The solar energy input is not distributed uniformly over the surface of the Earth, and the energy is redistributed by both horizontal and vertical motions in the atmosphere and oceans, which affect the relation between the effective temperature and the global average surface temperature.

Where we began by considering the energy equilibrium of the entire Earth, we have been led to examine the balance for the surface and atmosphere separately, still averaged over the globe. Later we shall consider the situation as a function of latitude, comparing say equator and poles. Finally we might look at a local situation. Thus the energy equilibrium at Hawaii for example, will depend not only on vertical energy flows-input from the Sun, long-wave radiation to and from the atmosphere, geothermal inputs (it is a volcanic "hot spot"), non-radiative energy losses, notably by evaporation; it will also include horizontal flows – ocean currents, trade winds, precipitation of water which came from elsewhere. Such a calculation can be made too for a space ship: then the only inputs and outflows are radiative (some non-thermal, as when an S-band transmitter sends messages back to Earth). However, local energy production, which may include the metabolism of food if the vessel is manned, as well as power production by fuel cells or nuclear sources, must be taken into account. Also, for interplanetary probes, the solar flux varies as the space ship's distance from the Sun changes.

The common feature of all these calculations is that they are energy *equilibrium* calculations; they assume balance between inputs and outflows of energy, and thus they assume that everything is constant. This assumption simplifies the calculations enormously. However, we have already remarked that perfect constancy is practically inexistent in nature, whether we consider very short time scales or very long ones. Real systems are at best only approximately in equilibrium. The climate system is one such. Now one way of envisaging change is to examine a series of slightly different equilibrium states, without worrying about the details of the transitions from one to the other. Many global climate calculations are carried out in this way. Thus the present global average surface temperature might be determined assuming equilibrium as the consequence of the present values of the solar flux, the global albedo, the infrared opacity of the atmosphere, and a non-radiative energy transport parameter. If new different values of one or more of these parameters are "plugged into" the calculation, which assumes equilibrium, a new global average surface temperature is computed. For example, a higher CO_2 content of the atmosphere implies a higher infrared opacity, which leads to a higher surface temperature, all other things being equal. The other things may not remain equal. The equilibrium calculation does not really examine how the climate changes as the CO_2 increases, because this is necessarily a nonequilibrium process. Many, although not all, global climate models are "linear", in that only one effect is considered at a time. In a linear model, for example, while a CO_2 increase leads to an increase in temperature, the possible effects of this temperature increase on the global albedo, or on the atmospheric CO_2 itself, are not considered. There is no feedback. In most real systems there is feedback. Nature is not linear.

These considerations now lead us to attempt a more precise definition of the term stability. Stability is not complete invariability. Let us recognize once and for all that such constancy is absent in nature, that change and fluctuation are everywhere, and that a real system of any sort is never perfectly in equilibrium. A particular system may however stay very close to equilibrium, with only small fluctuating imbalances or perturbations, if the equilibrium is stable. This will generally be the case if any departure from equilibrium is promptly (but sufficiently gently) corrected, i.e. if the system has negative feedback (but not too strong) to perturbations. By contrast, if the system responds with positive feedback, the departure from equilibrium is amplified, and we have an unstable situation. The two cases are illustrated in Fig. 5-9. Note that excessive negative feedback can lead to growing oscillations; we have not illustrated this case.

We can find examples of both negative and positive feedback in various subsystems of the climate system, as shown in Fig. 5-10. Thus, where heating by sunlight is concerned, the cycle of solar heating, evaporation and cloud formation is a factor of stability. On the other hand; because snow reflects sunlight so well, and persists when temperatures stay low, we have the well-known ice-albedo instability. Indeed, simple climate models computed by the Soviet climatologist Mikhail Budyko, and the American William Sellers, suggest that the present climate should be highly unstable, and that a perturbation should lead either to a completely ice-free world (which we recall has been the most common situation over most of the past 600 Myr at least), or to a completely ice-covered world (which seems never to have occurred). Of course the difficulty in the study of climate is that the geosystem is extremely complex, involving many different processes interacting with one another. Because these processes operate at different speeds, there are time lags in the response of some elements of the geosystem to changes in climate factors, and departures from equilibrium can persist. Thus the oceans "remember" the climate of the past much longer than do land areas. Some of the geosystem's processes

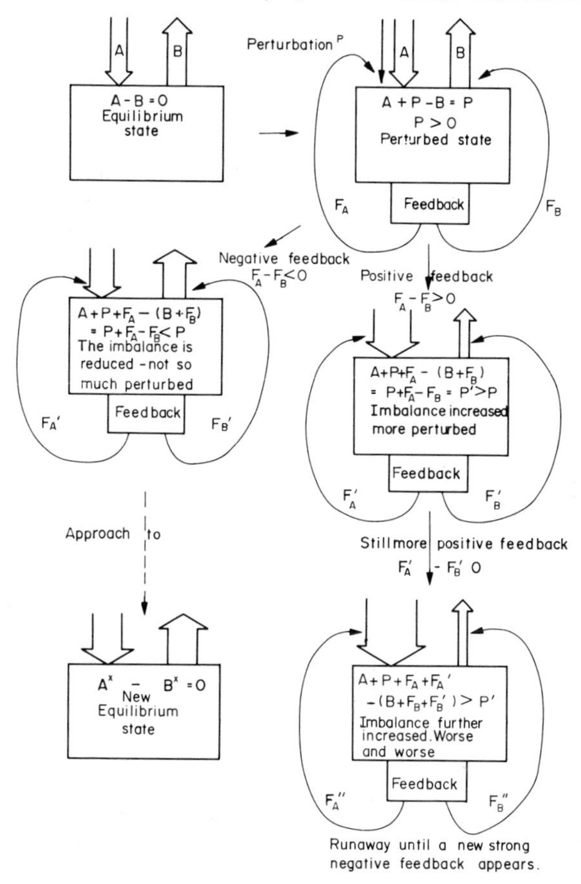

Figure 5-9. Positive and negative feedback.

contribute stability since they have negative feedback to perturbations of climate, while others, like the ice-albedo relation, are factors of instability. Whether instability or stability dominates depends on the relative importances of these different processes, as well as on the size of the external perturbations. It may be that the system corrects itself up to a certain point, but that a really big perturbation, such as a period of unusually violent volcanic activity, can trigger a "new" instability. We are

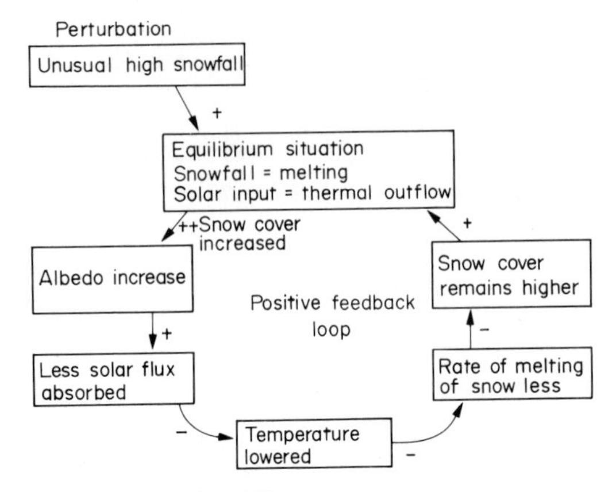

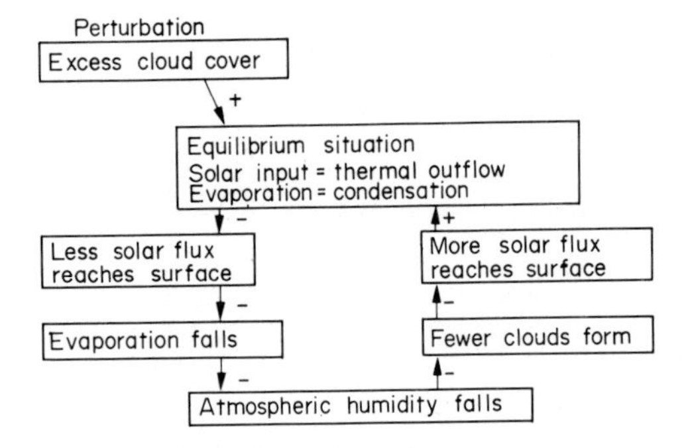

Figure 5-10. Examples of positive and negative feedback in the climate system. Note that in each case one could change the sign of the perturbation. The point is that in the positive feedback loop the sign is the same at the end of the loop as at the beginning, whereas it changes in the case of negative feedback. Note also that there are other aspects of the water cycle (notably with regard to infrared radiation) which may involve positive feedback.

certainly not now in a position to make precise predictions of future climate with confidence.

We may be able to learn something from comparison of the Earth with other planets. The planet Venus is similar to the Earth in size and mass, but its "climate" is extremely different. Although it is closer to the Sun, so that the solar flux is about twice as high as on Earth, its albedo is so high that its effective temperature comes out as 244°K, lower than the Earth's. Nevertheless, its surface temperature is far higher (700°K!) essentially because the CO_2 content and infrared opacity of its atmosphere are enormous. Ichtiaque Rasool and Catherine de Bergh of NASA's Goddard Institute for Space Studies in New York ascribed this situation to a "runaway greenhouse effect" early in the planet's history, i.e. to strong positive feedback of the atmospheric CO_2 (and H_2O) on the processes releasing these gases from the surface of the planet.[8] This might also have operated on the early Earth, but here an even stronger negative feedback, linked to the formation of liquid water and carbonates, must have arrested the process. The essential difference may have been the level of the solar flux, outside the primitive atmospheres of these planets, which was of course higher for Venus.

I hope that the concepts of stability and instability, as associated with negative and positive feedback respectively, are somewhat clearer now. If we think of the climate system as constituted principally by the Earth's ocean and atmospheres, and some aspects of the surface, then clearly there are factors like the solar radiation, the orbital parameters, solid earth processes like continental drift and volcanic activity, or human processes like the burning of fossil fuels, which are "external" to this system. When these external factors vary, they trigger responses in the climate system. We say that the system, and thus the Earth's climate, is stable, if for small variations in the external factors, no dramatic changes result; this will be the case if no dominant positive feedback process operates. On the other hand, if the system is unstable, very small fluctuations in the external factors may still trigger major changes; indeed the climate may vary significantly without any change at all in the external factors, purely as a result of positive feedback reactions among the "internal" processes. We do not know which is the case. In any event, major changes in the external factors necessarily have an impact on climate.

6

Sun and Earth

> But in the center of all resides the Sun. Who, indeed, in this most magnificent temple would put the light in another, or in a better place than that one wherefrom it could at the same time illuminate the whole of it? Therefore it is not improperly that some people call it the lamp of the world, others its mind, others its ruler, Trismegistus calls it the visible God, Sophocles' Electra, the All-Seeing. Thus, assuredly, as residing in the royal see the Sun governs the surrounding family of the stars.
>
> Nicolaus Copernicus,[1] 1543

OF the external factors governing climate on Earth, the most important by far is the Sun. It is the flow of radiant energy from the Sun which drives all atmospheric and oceanic motions. As already mentioned here, this energy flow is on the average over 1000 times greater than all other energy flows on Earth. Locally, for example during the winter in the Leningrad area, this may not be the case. The contribution from fossil fuel burning is important there. However, it should be noted that such combustion of fossil fuels, which constitutes most human power production today, is simply release of solar energy stored in biomass in the past. Only nuclear, geothermal and tidal power are not based on solar energy.

What is the source of the solar energy output? We have already mentioned that most astrophysicists believe it to be the result of thermonuclear fusion — the conversion of hydrogen to helium — in the central core of the Sun. Calculation of "models" of the Sun's internal structure accounts for the Sun's luminosity rather well, in consistency with its mass, radius and chemical composition. There is, however, one serious problem which remains unresolved. The "best" model of the Sun predicts that neutrinos should be released as part of the nuclear reactions in the core, and since these neutrinos are only affected by the weak interaction, most of them should escape from the Sun without any further interaction with solar matter. Some of them will pass through the Earth. Detecting them is obviously extremely difficult, since detection relies on interaction with

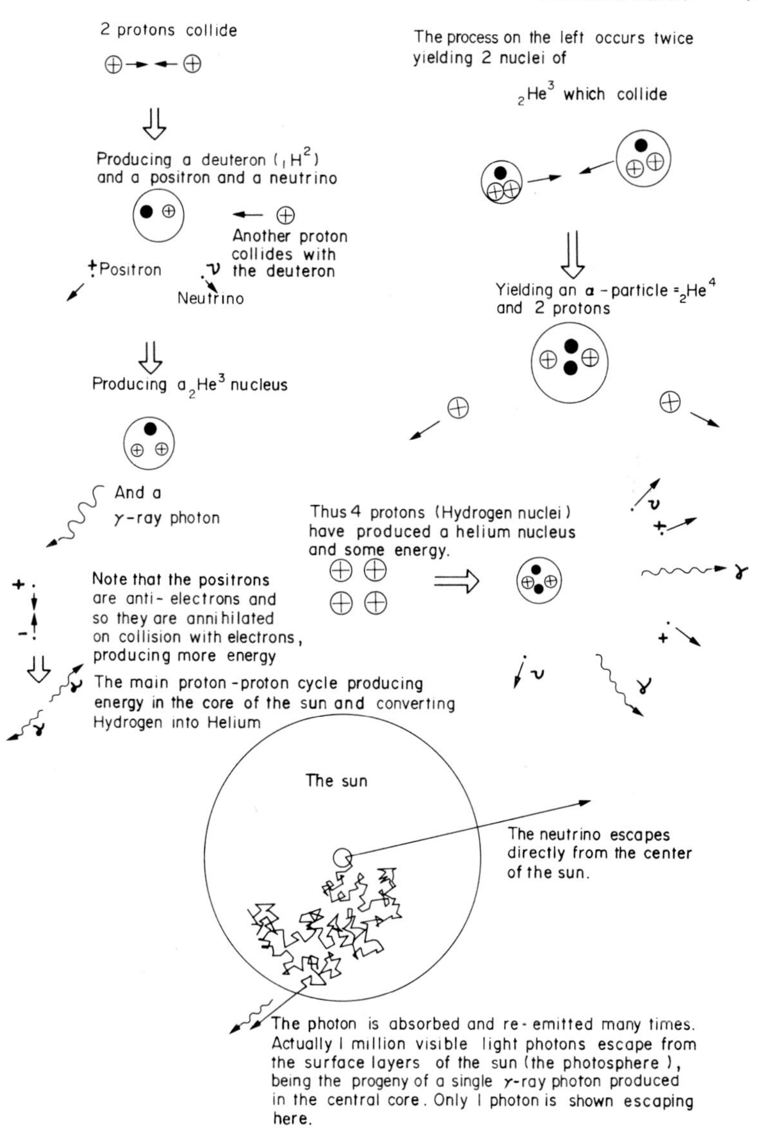

Figure 6-1. Production of energy in the central core of the Sun and its escape.

matter, but R. Davis of Brookhaven Laboratories in the United States has set up a neutrino detector consisting of a huge tank of cleaning fluid (carbon tetrachloride, the chlorine being the important nucleus), in a mine nearly 2 kilometers underground in South Dakota. With this he should be able to count solar neutrinos. The puzzle is that hardly any have been detected, far fewer than predicted by the models.[2] Thus we have some doubts about the solar model, or about the behavior of neutrinos, but still most astrophysicists believe, or hope, that it is only a detail, and not the whole picture, that is wrong. I shall proceed assuming this to be so.

In the generally accepted view of the Sun, following a "birth" phase about which we still know very little, we believe that hydrogen "burning" has been going on smoothly over the past 4 eons (billion years), with only a very gradual increase of the solar luminosity, corresponding for the most part to very slow dilation of the outer layers as the central core is enriched in helium. This implies that say 3 eons ago, the solar flux must have been somewhat lower than now, and if the global climate was then as warm as it is now, there must have been some factor compensating for this. Understanding the climate of the remote past involves understanding both the evolution of the Sun and changes in the Earth's atmosphere. However, over intervals of millions or even tens of millions of years, the solar luminosity can be taken as very nearly constant, at least according to the usual model. This would seem to disconnect the Sun from the ice ages.

Some astrophysicists, notably the Estonian-Irish astronomer Ernst Öpik, and more recently Douglas Gough in Cambridge, England, have suggested that the great ice ages every hundred million years or so, are caused by variation of the solar luminosity.[3] Dilke and Gough also link this to the missing solar neutrinos. In their theory, the enrichment of the central core does not proceed perfectly smoothly, but periodically involves partial mixing with unenriched material from farther out. In these phases, the rate of energy production in the core drops slightly, and this is immediately (i.e. in 8 minutes) manifest at the Earth by a very large drop in the neutrino flux from the Sun (presumably the present situation). The energy we receive from the Sun is not in neutrinos, however, but in photons emitted by the outer layers, and while these are the "progeny" of the X-ray photons produced at the center, it takes typically ten thousand years for the drop in central energy production to "appear" at the surface. In this view, today's missing neutrinos herald tomorrow's ice age. However, many objections have been raised against this view, and it is not certain whether they can all be met.

Let us put these worries aside, and examine how the energy produced in the core emerges from the Sun. Through much of the interior of the Sun, this is a matter of (about 10^{22}) successive absorptions and emissions of photons in a very roundabout "random walk" out from the center. This accounts for the ten-thousand-year delay compared to the direct travel time of 2.3 seconds from the center to surface, photons travelling at the speed of light. Inside the Sun, say two-thirds of the way down to the center, the temperature may be very high, and enormous amounts of radiation, going in all directions, are present. However, because the temperature does decrease outward, there is a slight excess of outward-bound over inward-bound photons, corresponding in what we call radiative equilibrium, to exactly the net flux needed to carry out the energy produced in the center. The average energy of a single photon is quite high deep in the interior where the temperature is of the order of a million degrees, but just as the temperature falls as we go out, we encounter more and more photons of lower and lower average energy. The photons that finally escape from the solar photosphere and reach us have energies typically in the range from 0.5 to 3 electron-volts, corresponding to visible radiation.

There is one complication of this picture which turns out to be quite important. Below the photospheric layers, but in the outer part of the Sun, there is a zone in which the stratification of temperature and density is *unstable*, in the sense that we gave to this term at the end of the preceding chapter. When a small mass of gas is slightly less dense (hotter) than its surroundings, it is buoyant and begins to rise, becoming still less dense relative to its surroundings. Thus this region responds with positive feedback to local perturbations of temperature and density. Motions are generated and amplified in the gas, with generally warmer masses ascending and cooler masses descending, and on the average, thermal energy is transported upward by these motions. The atmosphere is no longer in radiative equilibrium here, since the part of the energy flux transported by these *convective* motions generally varies with height, with complementary variation of the part of the energy flux transported by radiation. Most of the thermal energy reappears as radiation at the top of this *convection zone*, but these motions also necessarily carry kinetic energy, and some of the motions penetrate into the stable atmospheric layers above. High-resolution photographs of the Sun show a fluctuating pattern of bright and dark regions, called the *granulation*, which is a consequence of these convective motions (see Fig. 6-3). Understanding the details of these motions is very difficult, but there is at least a strong suspicion that they are related in various ways to the phenomena

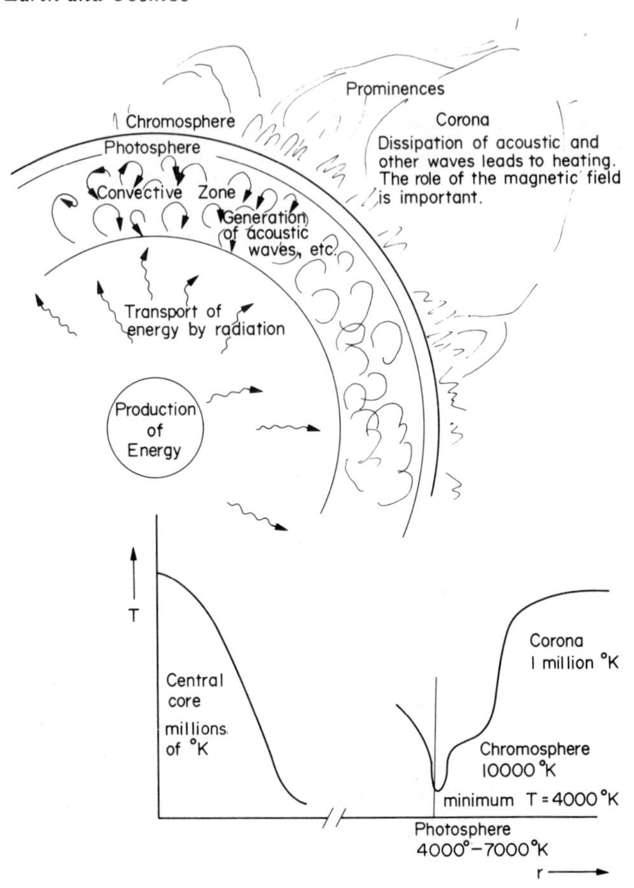

Figure 6-2. Modes of energy transport and the temperature structure of the Sun.

of solar magnetism and solar activity, through interaction with the rotation and possible vibrations of the Sun. They are thus probably indirectly responsible for much of the solar ultraviolet and shorter-wavelength radiation, for the solar wind, and for the energetic particles (solar cosmic rays) occasionally emitted by the Sun.

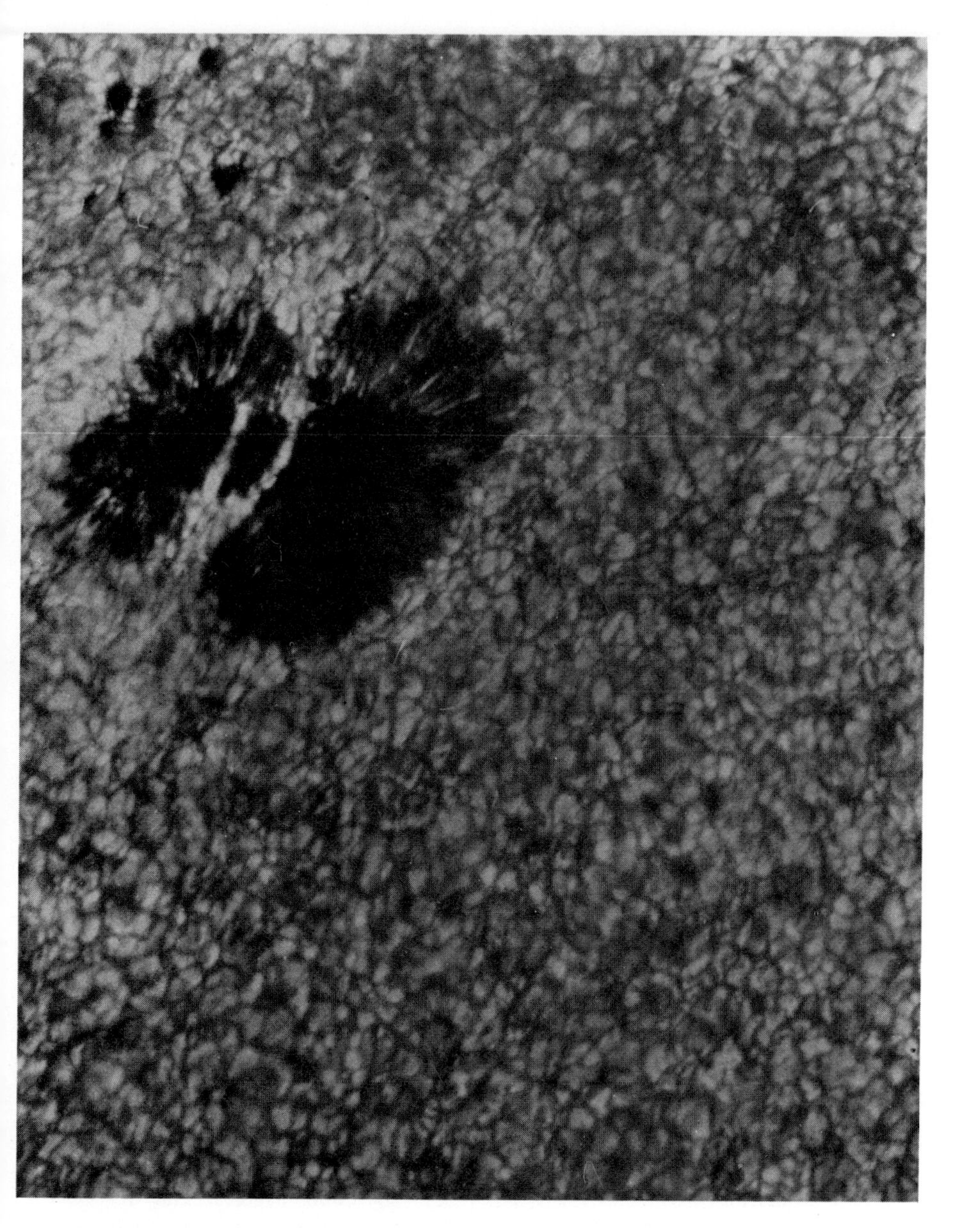

Figure 6-3. The surface of the Sun: granulation and a sunspot group. The granules discernible here are typically 1000 km across, and are caused by convective motions deeper in the solar atmosphere. The spots are several thousand kilometers across, rather larger than the Earth. White light photograph taken 3 June 1971 at the solar tower telescope of the Observatoire de Meudon.

Let us see how this can be. Energy flows out from the center of the Sun, and the temperature decreases outward, until we reach the photospheric layers from which most of photons escape. This is reasonable: heat flows from hot to cold. However, beyond the photosphere, observations show that the temperature begins to *increase* outward, in the *chromosphere*, reaching values over a million degrees in the extended tenuous hot atmosphere called the *corona.* These outer regions of the Sun's atmosphere were first detected by eye during total eclipses, but with appropriate techniques they can be studied continuously. Since temperatures are high, photons of high energy, i.e. of short wavelength, (UV, X-rays) are emitted. As we mentioned in the second chapter, such photons interact rather easily with matter; thus the solar photosphere is extremely opaque to such radiation and there is no chance of UV and X-radiation emerging from the deep hot layers of the Sun. When we "look at" the Sun (from space) in the ultraviolet, and X-ray domains, we "see" only the outermost regions.

The heating of these outer regions remains problematical. The energy probably comes from the dissipation of various types of waves — acoustic waves becoming shocks, hydromagnetic waves, gravity waves — these waves being generated by the turbulent motions in the convective zone, or linked to other types of instabilities. When we study the chromosphere and corona, we find extremely rich and at times rapidly varying structures at all scales, correlated with a complicated magnetic field structure. At times, extremely violent events — the solar flares — take place, in which considerable amounts of energy are released in a very short time, in the form of X-rays, UV and sometimes even visible light emission, as well as energetic streams of protons and electrons. It is believed that this is the result of sudden release of energy stored over extended periods of time by particular arrangements of the magnetic field, which finally become unstable and untwist themselves, but the details have not been worked out in a satisfactory way. The magnetic field itself is attributed to the interaction of the Sun's *differential* (i.e. latitude-dependent) rotation with the motions in the convective zone, where much of the hydrogen is ionized, but there too the theory is not very well established.

Actually all of the above chromospheric and coronal phenomena vary with the solar activity cycle, which was first discovered in connection with the *sunspots*. These appear as "dark" spots on the visible disk of the Sun, and in fact they are sharply limited regions in the photosphere where temperatures are depressed by 1000 to 2000 degrees, and where extremely strong magnetic fields are found. Apart from ancient naked-eye observations (mostly Chinese) of very large sunspot groups, sunspots were first described by Galileo in 1610, on the basis of telescopic observation. The well-known 11-year sunspot cycle was only identified in the middle of the last century, and generally accepted following the work of the Swiss astronomer Wolf. The 11-year rise and fall in the number of sunspots is only one aspect of the solar activity cycle, which includes many of the chromospheric and coronal phenomena mentioned above. The sunspots are associated with so-called active regions from which most of the solar short-wavelength radiation is emitted, and flares occur mainly in complex spot groups, where the strong magnetic field structure becomes unstable. Pairs of spots usually behave like the poles of a bar magnet, with a systematic polarity relation between leading and following spots in a given hemisphere of the Sun. This polarity relation is opposite in the two solar hemispheres, and it reverses from one cycle to the next, so that one should really speak of a 22-year solar magnetic cycle. The solar activity cycle is closely correlated with various phenomena on Earth, such as aurorae and fluctuations in the Earth's magnetic field, as we shall explain shortly. Many other correlations, with phenomena as diverse as rainfall, crop yields, the stock market, epidemics and wars, have been proposed; some of these manifestations of "cyclomania" can hardly be taken seriously; others are plausible but require critical examination of the statistics, and would be much more credible if detailed physical mechanisms could be proposed to account for them. Research in this area is being actively pursued in many places.

It has often been assumed that this 11-year activity cycle is just another aspect of a nearly constant Sun, repeating more or less regularly for the last few eons, and just as likely to continue in the future. This does not seem to be justified. Following Galileo, and despite the intense interest of astronomers, hardly any sunspots were observed between 1645 and 1715, and this does not seem to be due to observational difficulties. This was noted by Maunder in England in 1894, and the historical record has recently been re-examined by J. Eddy in Boulder, Colorado,[4] who gives further evidence that this "Maunder minimum" was quite real, and that many other signs point to the low level of solar activity during this period. Neither chromosphere nor spectacular coronal streamers were noted during total eclipses, and aurorae seem to have been seen very rarely. Furthermore, the carbon-14 isotope, whose production depends on cosmic ray particles impacting CO_2 molecules in the upper atmosphere, is anomalously abundant in tree rings corresponding to this period, and this can be understood as the result of additional galactic cosmic rays reaching the Earth with the solar wind "turned down" during an inactive period. Finally Eddy notes the coincidence of the Maunder minimum with

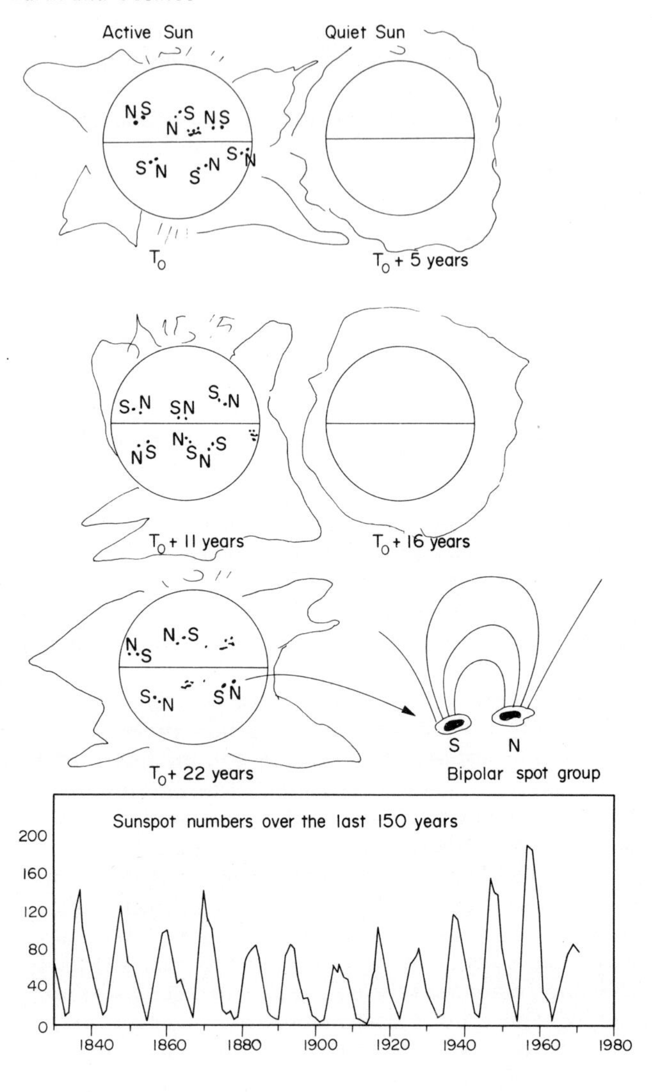

Figure 6-4. The sunspot cycle. Note that the period of the magnetic polarity cycle is 22 years. The horizontal line represents the solar equator.

the Little Ice Age, which we described in the previous chapter. He speculates that the C^{14} anomalies found for earlier times, notably the "Grand Maximum" about A.D. 1100–1200 which correlates well with the Little Optimum Period, point to other variations in the level of solar activity. He also has noted evidence that the differential rotation of the surface layers of the Sun may have been significantly changed at the start of the Maunder minimum.

We see that many aspects of the Sun remain poorly understood. While generally the Sun appears to us to be constant, we really have only a very short time base of precise observations, and we know that certain aspects do vary, at times dramatically. Let us again recall that observation of the complete solar electromagnetic spectrum has been possible only since the space age began, and has become complete only with the launching of the orbiting solar observatories. Long-term monitoring, with careful calibration, is necessary. Also, even today, we have never observed the Sun from over its poles, far from the plane of the Earth's orbit. A joint NASA-ESA "Solar Polar" mission is planned for the future, and it will give us a new perspective on the high-latitude regions of the Sun, and in particular on latitude-dependent aspects of solar activity.

From the ground, the most ambitious and systematic effort to monitor the solar flux, integrated over all wavelengths, was that of Abbot, of the Smithsonian Institution, who claimed to have found significant variability (a few %) of the "solar constant", correlated with weather changes. These measurements were made over the first half of this century. However, much of this supposed variability has since been explained (by others) as the result of inevitable errors of measurement, or of problems related to deducing the value of the solar flux outside the atmosphere, from ground-based observations. The "Solar constant" is near 1370 watts per square meter, but we really do not know whether the absolute value is closer to 1350 or 1390, nor do we know what it was, say, in 1700.

In the preceding chapter we gave a caricatural picture of what happens to the solar flux when it is intercepted by the Earth. Now let us examine in some detail the interaction between this radiation and the matter of the Earth's atmosphere. Our atmosphere is made up mostly of N_2 and O_2 molecules, and the interaction between these molecules and the visible and near infrared radiation which constitutes most of the solar energy flux, is relatively slight. Only a small fraction of the photons in this spectral domain, except at particular energies corresponding to spectral lines, is *absorbed* by the nitrogen and oxygen molecules. On the other hand, a substantial fraction are *scattered*, i.e. the direction of the photons is changed. This scattering by the molecules was well described by Lord

Rayleigh and so is often called Rayleigh scattering; photons of short wavelength (but still long compared to the sizes of molecules) are much more likely to be scattered. In the visible domain, this means that blue light is more likely to be scattered than red. The result is that the Sun appears somewhat yellower or redder than it would from space, and the sky is blue. The blue light of the sky is simply made up of photons, preferentially blue, removed from the direct solar beam, and scattered into different directions. The "white" light in a beam of sunlight is divided into one man's red sunset, and another man's blue sky. Some of the light is scattered upward, and this contributes to the global albedo. Indeed we can see the sky from above; in a commercial jet at an altitude of 9000 meters we are already above most of the atmosphere, and looking down we see the light, mostly blue, scattered by the atmospheric molecules, superimposed on light reflected from the ground. The beauty of our "blue planet" from space has been described and photographed by the astronauts.

All of this assumes a pure blue sky. There also are solid particles and liquid droplets — *aerosols* — in suspension at various altitudes in the atmosphere. They can both scatter and absorb light in the visible and near infrared domain. Usually it is assumed that their major role is in increasing the global albedo, and thus that their overall effect on the Earth's climate is a cooling one, but this is by no means well established, and the opposite has been suggested too. The optical properties of aerosols depend on their sizes as well as on what they are made of. Generally they do not favor the short wavelengths as much as the smaller molecules do, mainly because they are larger than all the wavelengths over the entire visible and near IR domain.

There also are clouds, which can both scatter and absorb light. They generally appear white, thus their reflectivity does not depend on wavelength in the visible domain. Clouds are by far the major contributors to the global albedo. Thus anything that affects mean cloudiness, as for example the pattern of global circulation of the atmosphere, the rate of evaporation, or the availability of solid particles which can trigger condensation of water vapor and thus the formation of clouds, can affect the global albedo. In certain areas previously known for many cloudless days, such as Denver, Colorado, an increase in cloudiness of the *cirrus* type, clearly triggered by exhaust from high-flying jet aircraft, has been noted. Whether such "jet cirrus" can have a measurable effect on the global albedo is not yet certain.[5]

Finally, some of the sunlight reaches the surface of the Earth, either in the direct solar beam, or in the diffuse light from the sky and clouds.

What happens then depends on both the direction of the beam and the nature of the surface. We have already mentioned the rather high albedo of ice and snow. Over the ocean, on the other hand, the albedo can be very low, particularly for vertical or near-vertical incidence of the beam, as is the case in the tropics. A major part of the solar energy flux is absorbed and injected into the geosystem in the tropical oceans. On land surfaces, a wide range of albedo is found, depending on the type of vegetation (if any), roughness, and of course the extent of any snow cover. On the average, over the whole globe, the Earth's surface reflects back upward about 14% of the visible light flux it receives. This reflected flux may again interact with the atmosphere on the way out. All in all, about 30% of the visible light from the Sun is reflected back to space, about a fifth of it is absorbed, and thus transformed into heat, at various levels of the atmosphere, and not quite half of it is absorbed and turned into heat at the Earth's surface, mainly in the tropical oceans. We shall examine the quantitative energy budget in more detail in the next chapter. There is little evidence at present for variations in the visible light solar flux (once the slightly changing Sun—Earth distance is taken into account), and the tiny variations suggested by ground-based observations may well be due to variations in the transmissivity of the Earth's atmosphere rather than to intrinsic solar variations.

The picture at the very short wavelengths — UV and X-rays — is very different; here substantial solar variations exist. What happens to these photons? The solar UV photons reaching the vicinity of the Earth originate in the outer solar atmosphere and not in the deeper layers, because they interact very easily with matter. For the same reason, these photons cannot penetrate deeply into the Earth's atmosphere, and we can only observe them from rockets or satellites, above the atmosphere. While these photons do not carry a large fraction of the total energy flux from the Sun, they are individually high-energy photons, and the effects they have on the upper atmosphere of the Earth are extremely important.

The very short wavelengths, up to about 100 nanometers, correspond to photon energies above 12 electron-volts, high enough to remove electrons from atoms and molecules of the upper atmosphere. This process is called *photoionization*, and some typical reactions are shown in Fig. 6-6. Some of the photon's energy is used to overcome the electromagnetic force binding the electron to the rest of the atom or molecule, for example 13.6 eV for the oxygen atom; the remaining energy appears in the motions of the resulting ion—electron pair, and thus contributes to raising the temperature of the gas. Of course a wide variety of reactions between ions, electrons, atoms, molecules and photons can take place. The chemistry of

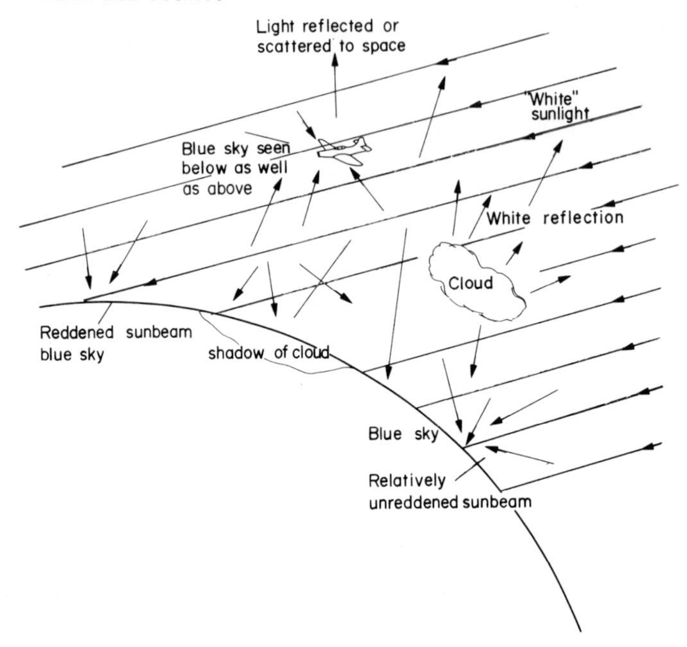

Figure 6-5. Scattering and reflection of sunlight in the Earth's atmosphere. The wavelengths involved go from roughly 330 to 1400 nm, i.e. the visible and the very near infrared.

the upper atmosphere is very complicated, and because photons play an essential role, we often speak of "photochemistry". For example, at certain altitudes, nitric oxide (NO) is an important source of electrons (Fig. 6-6) when it absorbs solar Lyman-α photons, but its production is extremely complex, as we shall see in chapter 12 when we discuss possible human impact on the ozone layer.

In the uppermost layers of the atmosphere, the temperatures reach very high values, and so this is called the *thermosphere*. Since the heating depends on the far ultraviolet photon flux from the Sun, it fluctuates with solar activity (see Fig. 6-7); temperatures also fall at night. When the temperature is highest, the increased agitation of the atoms and molecules can support a more extended atmosphere. It was the unexpectedly rapid rise of solar activity in 1976—78 that led to the rapid decay of the orbit of Skylab, by atmospheric friction. The ion and electron densities also vary

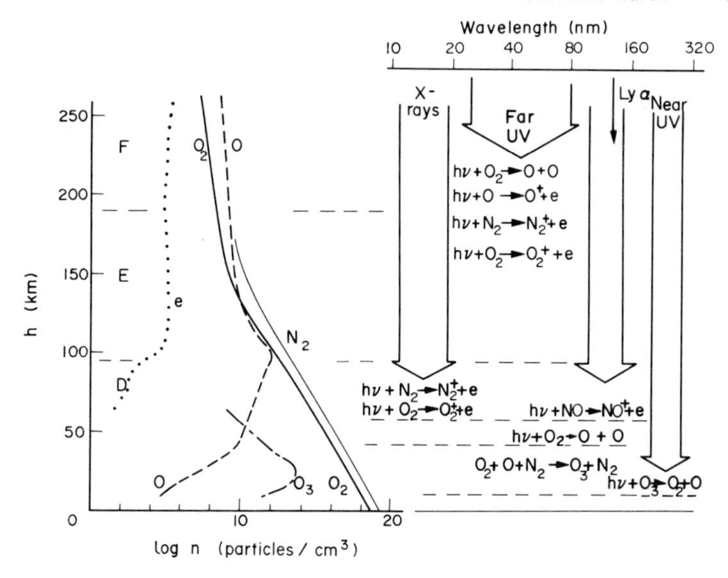

Figure 6-6. The structure of the ionosphere and ozonosphere. The logarithm of the particle densities is shown as a function of height in the atmosphere. To the right, some typical photochemical reactions are shown, corresponding to the absorption of incoming solar photons (represented by hν) in the wavelength range 10—160 nm and up to 320 nm where ozone is involved.

in response to the flux of energetic solar photons. In the ionosphere, between sixty and several hundred kilometers altitude, electric currents can flow and generate magnetic fields. Thus even though most of the gas remains neutral, its behavior is significantly changed. One of the most important properties of the ionosphere is that it refracts and reflects radio waves in certain wavelength domains, notably from 15 to 50 meters. These so-called short waves are indeed short compared to other often used radio waves, but they are very long compared even to much of the modern ultra-short-wave or ultra-high-frequency radio domains, not to mention the infrared or visible regions of the electromagnetic spectrum. In the past, this property was extremely useful, because it made intercontinental radio communications possible. It still is used today, but communications satellites offer much more reliable conditions of transmission at other frequencies which are unaffected by the ionosphere. The iono-

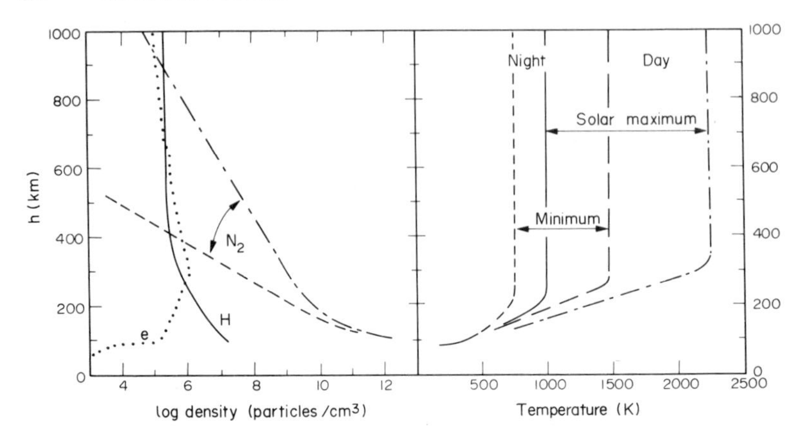

Figure 6-7. The thermosphere. Note how temperature depends on solar activity and varies from night to day, entraining similar variations in atmospheric density.

sphere is certainly a nuisance as far as radioastronomy is concerned, except insofar as powerful instruments financed by agencies interested in ionospheric research (notably the military) have been available for radioastronomy. The ionosphere remains a sensitive detector of various aspects of solar activity, and may be an important link between solar fluctuations and terrestrial weather fluctuations below. It certainly affects the magnetic field at the Earth's surface.

The structure of the ionosphere, and notably the peaking of the ion and electron densities near 300 km, depends on two opposing tendencies. Because of gravitation, gas density as a whole decreases with altitude. The effect is not very strong for the light atoms of hydrogen and helium, but it dominates the behavior of oxygen and nitrogen, (see Figs. 6-6 and 6-7), and the electrons come mainly from these constituents. On the other hand, the photoionization process itself both requires energetic photons and destroys them; below 100 km, hardly any such photons remain. These processes were first described by Sydney Chapman in England, some 50 years ago, and so we call the characteristic peaking of the density at a certain height a Chapman profile. An analogous situation accounts for the peaking of the ozone density between 15 and 50 km altitude. Here the photons involved are not energetic enough to ionize the gas. However, wavelengths less than 242 nm, which reach these layers, can *dissociate* the oxygen molecule, producing free oxygen atoms. The gas densities in these layers are sufficiently high so that triple collisions are not extraordinarily rare, and they make possible the formation of ozone, for example by the reaction:

$$O_2 + O + N_2 \rightarrow O_3 + N_2$$

Of course, two-body collisions of O_2 and O are much more frequent, even higher in the atmosphere, but they cannot produce ozone because energy and momentum could not both be preserved in such a case. Ozone is important not because there is so much of it, but because it can be dissociated by photons with wavelengths as high as 319 nm;

$$O_3 + \text{near uv photon} \rightarrow O_2 + O$$

This process is responsible for the sharp cutoff near 300 nm in the solar ultraviolet radiation reaching the ground. Although the total energy involved is small compared to the solar flux in the visible and near infrared regions of the spectrum, it is large enough to lead to considerable heating in these atmospheric layers. As a result the temperature increases outward from about 15 to 50 km, in contrast to the situation below, where temperature decreases with height, and where the principal heating is at the Earth's surface. The lower layers are often unstable, they are the domain of weather, and generally of motions that keep them well mixed; this region is called the *troposphere*. A temperature minimum of about 180°K (−93°C) occurs at what is called the tropopause, at 15 km, and from there to about the 50 km level, called the stratopause, the temperature rises to about 270°K. Since the temperature does increase outward in this ozone layer, convective instability cannot occur, and the stratification of this region should be stable: it is called *stratosphere*. From the stratopause out to the mesopause at about 100 km altitude, the temperature again falls; this region is called the mesosphere. Further out, the processes that account for the ionization also produce heating, as already mentioned; in this *thermosphere* the temperature again increases outward.

The electromagnetic forces acting on the ions and electrons of the outer atmosphere, due to the existence of the Earth's magnetic field, also must be considered. They have some measurable effects at altitudes as low as 100 km, in the E region, and the magnetic field that we measure on the ground is in turn affected by what happens in the ionosphere. Because the daylit side of the Earth heats up, the atmosphere as a whole moves with a daily rhythm. The ionosphere is pushed up and down by these pulsations, and these motions of the layers carrying free electric charges act like the movement of a dynamo, inducing voltages and a current which in turn induces small but definitely measurable magnetic fields. This appears as a cyclical fluctuation of the Earth's magnetic field. This is only one of many effects of the day/night cycle of sunlight interacting with the Earth's atmosphere and magnetic field. Finally, as we go further up in the atmosphere, the gas pressure decreases, as does the gravitational attraction of the Earth, so that where ions and electrons are concerned, the forces exerted by the Earth's magnetic field become completely dominant. The region in which these forces dominate is called the *magnetosphere*. The extent of this region is determined, on the one hand, by the strength of the Earth's magnetic field, and on the other, by the strength of what is called the solar wind, a flow of charged particles (mostly protons, electrons and helium nuclei), having its origin in the solar corona, and moving at speeds of several hundred kilometers per

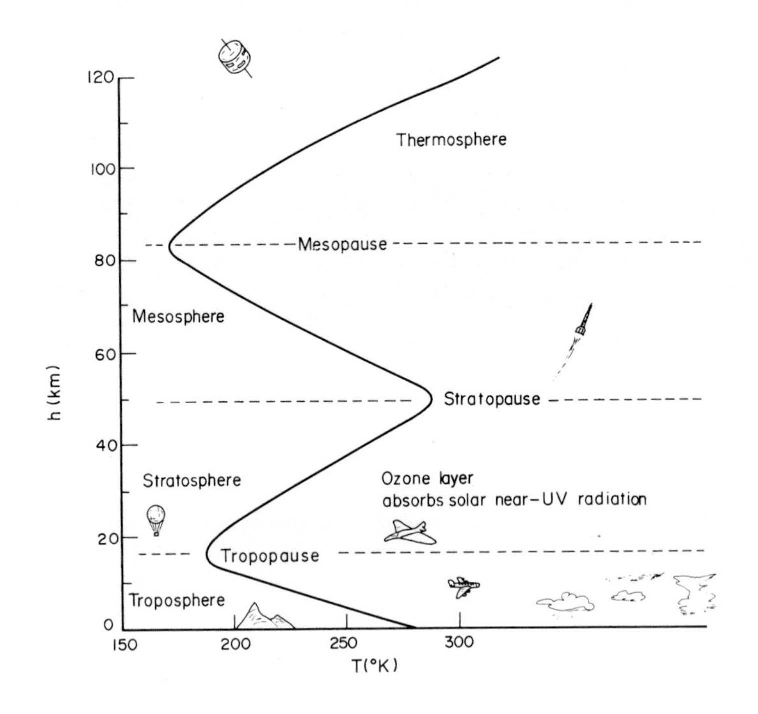

Figure 6-8. The thermal structure of the atmosphere up to 120 km.

second past the Earth's orbit, and having its origin in the solar corona. Both the Earth's magnetic field and the solar wind have been observed to vary on short time scales (100 years and less, even much less). We also know that the Earth's magnetic field has reversed itself many times over the past several hundred million years, probably passing each time at least briefly through a phase of essentially zero strength. During such a phase there could be no magnetosphere, and the solar wind would have direct effects on the upper atmosphere, nearly sweeping over the Earth. This is the case for the planet Venus, which rotates very slowly and has no substantial magnetic field.

Since the Earth *does* have a magnetic field, the solar wind, consisting of charged particles, is deflected. As a result, the magnetic lines of force are deformed, in such a way that the kinetic pressure of the solar wind is balanced by the magnetic pressure of the lines of force squeezed closer to the Earth. The interface is neither gentle nor smooth; as revealed by satellite measurements, the solar wind encounters and is deflected by a "magnetic shock" at a distance of about 12 Earth radii, where the value of the magnetic field suddenly changes from its interplanetary level to a higher fluctuating level. This magnetically turbulent region inside the magnetic shock gives way, at the magnetopause at about 10 Earth radii, to the region dominated by the terrestrial magnetic field. These figures apply to the "bow" of the magnetosphere, the region facing the Sun. If there were no solar wind, the Earth's magnetic field would be roughly axisymmetric, but with the solar wind streaming by, it appears stretched on the "lee" side and compressed on the "windward" side. Indeed certain magnetic lines of force extend out to at least 30 Earth radii, halfway to the Moon's orbit (in the antisolar direction), forming a long tail along a neutral sheet, and the Moon occasionally will cross this magnetic tail. For comparison, the planet Jupiter has an even more spectacular magnetosphere, through which one of its large Galilean satellites, Io, passes regularly. Saturn also must have a magnetosphere, and it is no doubt significant that the Earth, Jupiter and Saturn all rotate rapidly, since we believe rotation to be directly related to the generation of the magnetic field.

The influence of the Earth's magnetic field explains many other aspects of the terrestrial environment. We have already described how the field produces the magnetosphere, a cavity in the solar wind. The energies of the solar wind particles, travelling at most at a thousand kilometers per second, are rather low, and cannot cross the magnetic field lines. Charged particles of much higher energies — the cosmic ray particles, travelling at very nearly the speed of light — have long been known, and indeed they are responsible (through collisions with atmospheric atoms and molecules) for part of the ionization in the Earth's upper atmosphere, notably in the D region. The cosmic ray flux observed on the ground is generally somewhat higher in the polar regions than near the equator, at least for the primary cosmic ray particles, as opposed to the secondaries which result from their interaction with the atmosphere; this is readily understood since these charged particles can reach the surface more easily following the magnetic lines of force, which converge on the magnetic poles. It is suspected that most cosmic rays are produced within the Galaxy, by supernova remnants; they are confined to the Galaxy by the galactic magnetic field, which changes their directions so much that they have an isotropic distribution before encountering the Earth's magnetic field. The energies of these particles cover a wide range, but low-energy particles do not reach the terrestrial environment, being excluded from the inner solar system by the rather weak solar magnetic field carried out with the solar wind. Thus it has been suggested that if the solar wind intensity were to fall, either as a result of a true solar variation, or as a result of the solar system entering a dust cloud, many more "low-energy" galactic cosmic rays might reach the Earth, while on the other hand the Earth's magnetosphere would be effectively isolated from solar cosmic rays.

One of the highlights of the first successful American satellite launching was the discovery in 1958 of the Van Allen belts of the magnetosphere, by the Explorer 1 satellite. These "radiation belts" are roughly doughnut-shaped (toroidal) regions in which there are large numbers of high-speed electrons, trapped by the Earth's magnetic field, spiralling back and forth around the field lines between two "mirror points", until ultimately they collide with atoms or molecules and lose energy. To the extent that the electron density in the Van Allen belts remains roughly constant, there must be approximate equilibrium between injection of energetic particles, and leakage from these regions. "Leakage" may take place where the belts are lowest, since collisions with ionospheric constituents are most likely there where the densities are higher. The origin and mode of injection of the high-speed particles are not fully understood. Many of the particles are believed to be of solar origin, but there is also some evidence of injection from the Earth's ionosphere itself. One possibility involves neutrons emitted in the nuclear reactions between cosmic ray particles and the nuclei of ionospheric constituents: once a neutron is removed from a nucleus, it is unstable and decays, yielding a proton and an electron. There have been cases of artificial injection, notably in the course of the notorious "Starfish" experiments of 1962, when the United States detonated nuclear weapons at high altitudes in tests of the effects of such explosions on military communications. An artificial radiation belt was created, which persisted for many months, and which caused damage to the instrumentation of scientific satellites passing through this belt. The high densities of energetic particles are a definite hazard not only to equipment but also to humans passing through the radiation belts. A similar and indeed much more important radiation belt surrounds Jupiter, and is responsible for that planet's strong nonthermal radio emission at decametric wavelengths, which moreover is strongly perturbed when the satellite Io passes through the belt. This type of radio noise arises when energetic electrons are moving in a magnetic field, as is the case in the Van Allen belts of the Earth and Jupiter, as well as in other larger objects in the universe, such as the Crab Nebula. It is also the case in the *synchrotron*, a physicists' research tool, and for this reason the process is called synchrotron emission. This kind of radio noise is quite different from *thermal* radio emission, which is simply the consequence of the much slower thermal motions of atoms, ions, electrons and molecules, and which in the case of Jupiter and the Earth dominates at shorter (< 1 m) wavelengths.

We have not yet said very much about the solar cosmic rays, the energetic particles emitted by the Sun. These are responsible for part of the ionization and heating of the Earth's upper atmosphere, which they enter principally at fairly high geomagnetic latitudes, following the magnetic lines of force. There, they also account for the aurorae, which have awed men in northern latitudes since the beginning of history. These lovely changing displays of colored light are usually observed only in subpolar regions, and their frequency follows the solar activity cycle. They consist of groups of emission lines in the red and blue domains of the spectrum, emitted by molecular nitrogen, as well as a green line due to atomic oxygen. Using observations made at different places, the height of the aurorae can be determined, and typically it is about 100 km. At these heights atomic oxygen is available in substantial amounts. Nevertheless, the densities are low compared to what is feasible in the laboratory, and this was finally recognized in 1924 as explaining why the green auroral line had never been seen in a laboratory spectrum of oxygen. Much of the auroral zone is very sparsely populated, so that ground-based sightings only give a fragmentary picture; satellite observations now give us a global view (Fig. 12-8).

The number of energetic particles emitted by the Sun varies considerably, not only over the 11-year activity cycle, but also on much shorter time scales. We have already mentioned the effect of the day/night rhythm on the magnetic field measured at a ground station. This regular diurnal oscillation, whose amplitude tends to follow the sunspot number, can be understood as the effect of the very short wavelength UV and X-ray

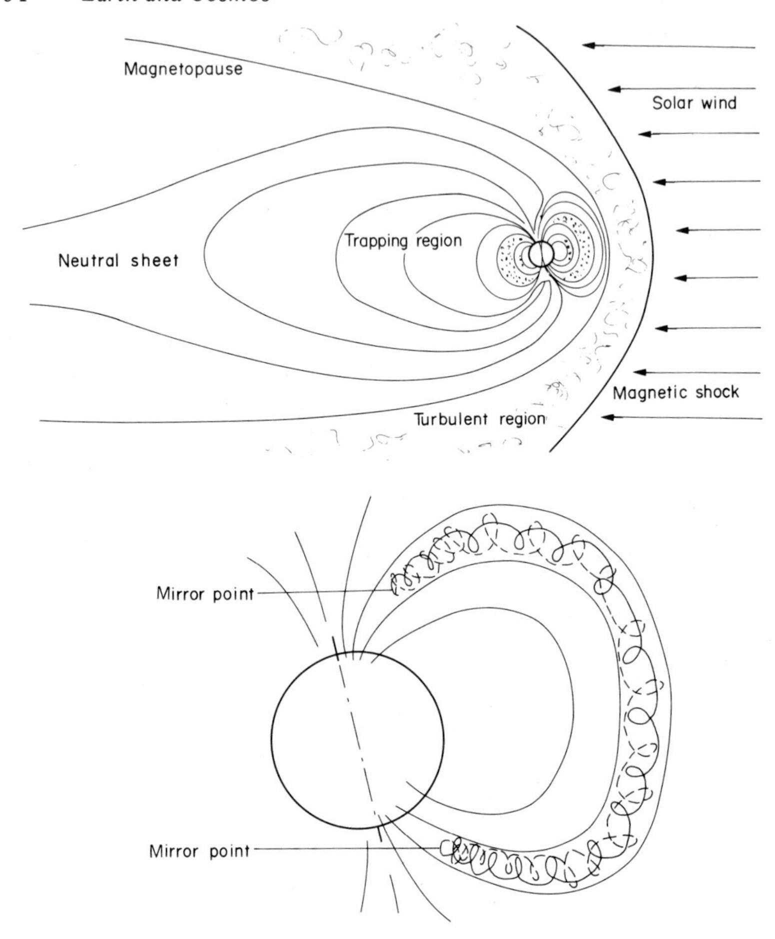

Figure 6-9. The Earth's magnetosphere. Above, note how the magnetic field lines are compressed on the "windward" side, and stretch out in a magnetic tail on the "lee" side. Below, the trapping of charged particles in the Van Allen belt by the magnetic field lines.

photons whose number depends on solar activity. Occasionally, however, there are fairly strong and rapid changes in the geomagnetic field, usually beginning very suddenly, then lasting several hours. These are called *magnetic storms*, and were first observed 200 years ago by the great German naturalist von Humboldt. Both the frequency and the intensity of magnetic storms are very well correlated with the solar activity cycle, as is the frequency of aurorae. Both of these phenomena occur on the night side of the Earth as well as on the day side, so they cannot be directly caused by solar photons. In some cases, a magnetic storm recurs regularly at 27-day intervals; this is the period of rotation of the Sun as seen from the Earth, and this suggests that whatever causes the storm is emitted by a limited region of the Sun. Sometimes then, but not always, the storm can be identified as the effect of a particularly active region, usually associated with a large and complex group of sunspots. At any rate, we believe that somehow the complex configuration of magnetic fields in an active region in the solar chromosphere and corona leads to strong acceleration of charged particles, mostly protons and electrons, so that a stream of these leaves the Sun, moving much faster than the solar wind. After travelling perhaps a few days along a path that generally is not at all straight, this jet impinges on the Earth's magnetosphere. The magnetosphere is then further compressed, leading to a sudden change in the magnetic field observed at various ground stations — the "sudden commencement" of the magnetic storm. The Van Allen belts are perturbed, various oscillatory motions are generated and currents flow, leading to further fluctuations of the Earth's magnetic field. Energetic electrons reach the 100 km levels in the auroral zone, possibly accelerated in the Earth's magnetic tail as a result of the various perturbations, and excite the aurorae.

These magnetic storms are by no means the most violent events involving Sun and Earth. Solar *flares* have far more dramatic effects. A flare involves an extremely violent release of energy in the outer layers of the solar atmosphere, believed to be the consequence of an instability in the magnetic field configuration in an active region. Protons and electrons, and other charged particles, are very strongly accelerated, to high energies. Some of these particles lose their energy by collisions with the gas in the chromosphere or corona, heating it to very high temperatures (10^7–10^8 K). While the region is very hot, it emits a burst of X-rays, multiplying the X-ray flux from the Sun by a very large factor, and lasting several minutes. As this plasma cools, and ions and electrons recombine, ultraviolet radiation is emitted with considerable enhancement lasting as much as an hour. Strong emission also appears in various spectral lines in the visible, and on rare occasions the flare is visible in white light. Since the flare involves both energetic electrons and magnetic fields, synchrotron emission is produced in the radio domain, and we observe this as a solar radio "burst",

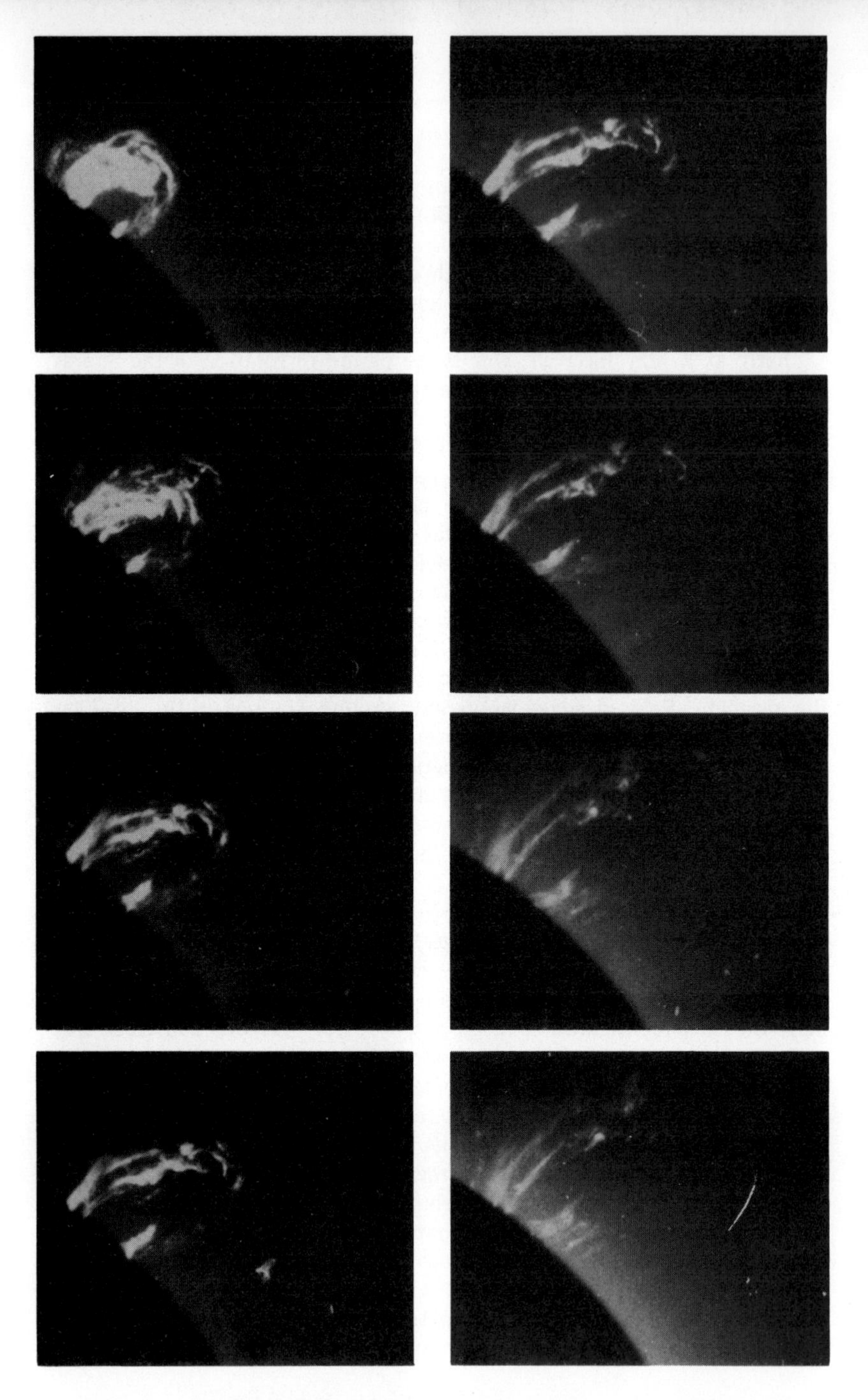

Figure 6-10. Solar flare, observed at the edge of the visible solar disk on 27 August 1977. Spatial scale: each image is 700 000 × 800 000 km. Time scale: 16 minutes from the first to the last of the images. Speeds at least as high as 280 km/s can be deduced. (Observatoire de Meudon).

which can be rather easily monitored from the ground, and localized on the Sun using interferometric methods. It should be noted, that just as in the Earth's ionosphere, conditions for transmission of radio waves in the partially ionized solar atmosphere depend on the frequency of the radiation and on the electron density in the gas. We can determine the evolution of the solar disturbance by studying how the radio spectrum we receive changes with time.

We have described in very summary fashion what occurs on the Sun during a flare. What happens on the Earth? The X-ray photons reach the day-time side of the Earth in 8 minutes, the light travel time, and trigger what is known as a *sudden ionospheric disturbance* (SID). The enhancement of ultraviolet flux tends to peak slightly later, but these photons also travel directly at the speed of light. They therefore produce immediate increases in the ionospheric electron density, particularly in the D region where the X-ray photons penetrate. This leads to strong absorption of radio signals, wreaking havoc in long-distance communication. Again currents are produced in the ionosphere, which induce fluctuation in the geomagnetic field, and these fluctuations can in turn induce currents in conductors on the ground, such as long-distance cables, occasionally damaging equipment or leading to the appearance of gibberish on teletypes.

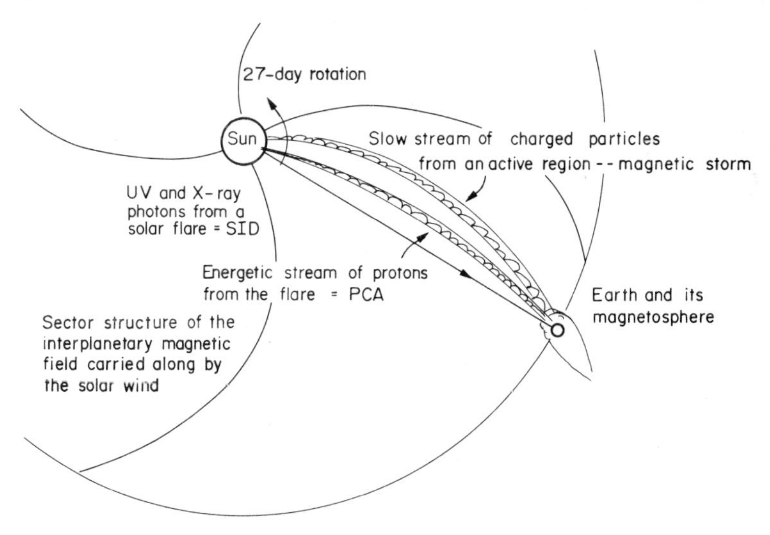

Figure 6-11. How different solar disturbances reach the Earth.

Some of the energetic particles accelerated in the flare escape the Sun, and they may reach the Earth some time later, having traveled not quite as fast as light, and having taken a somewhat roundabout route, depending on the interplanetary magnetic field. When very energetic particles including protons are involved, they may reach the Earth in as little as 4 hours, and penetrate to the D region at polar latitudes. There they substantially enhance the electron density, leading to what is called a *Polar Cap Absorption* (PCA) of radio waves. They also deposit considerable energy there, heating the gas. Finally there is evidence that by producing large quantities of nitric oxide (NO) they trigger large-scale reductions of the "odd-oxygen" ozone content. These effects are limited to the subpolar regions by the Earth's magnetic field; in its absence the entire atmosphere might be affected. Note that this is an extremely powerful event. More commonly, the flare particles take about a day to reach the Earth, and then they cause a magnetic storm and aurorae as indicated earlier, mostly indirectly through their perturbation of the ionosphere, rather than directly as in the PCA events.

It should be obvious that these solar events have an impact on many human activities on Earth. We have mentioned the problems related to telecommunications, which have stimulated international cooperation in studying these world-wide phenomena; they have also led to intense research by national military establishments. Considerable effort has been brought to bear on the problem of predicting solar flares, and significant progress has been made in devising synoptic methods. The importance of this has grown now that humans frequently travel (in Concorde, or in space laboratories) at altitudes where they are directly exposed to solar flare particles. There also are suggestions that such particles may influence weather, although it is not easy to understand exactly how. Both Joseph King at the Appleton Laboratory in Slough, England, and Walter Orr Roberts and his co-workers in Colorado have noted correlations between North Atlantic pressure troughs and changes in geomagnetic activity which presumably are related to solar activity, through the structure of the interplanetary magnetic field. Also, the cycle of droughts on the American Great Plains appears to be correlated with the 22-year solar magnetic cycle. Astronomers may yet have something to say about weather.[6]

In all of the solar-terrestrial relations we have discussed here, a great deal of symmetry appears between processes occurring on the Sun and on the Earth, even though these bodies are radically different. The outer regions of the solar atmosphere, which are anomalously hot because of the dissipation of various types of waves, emit the ultraviolet photons which are absorbed in the outer regions of the Earth's atmosphere, ionizing

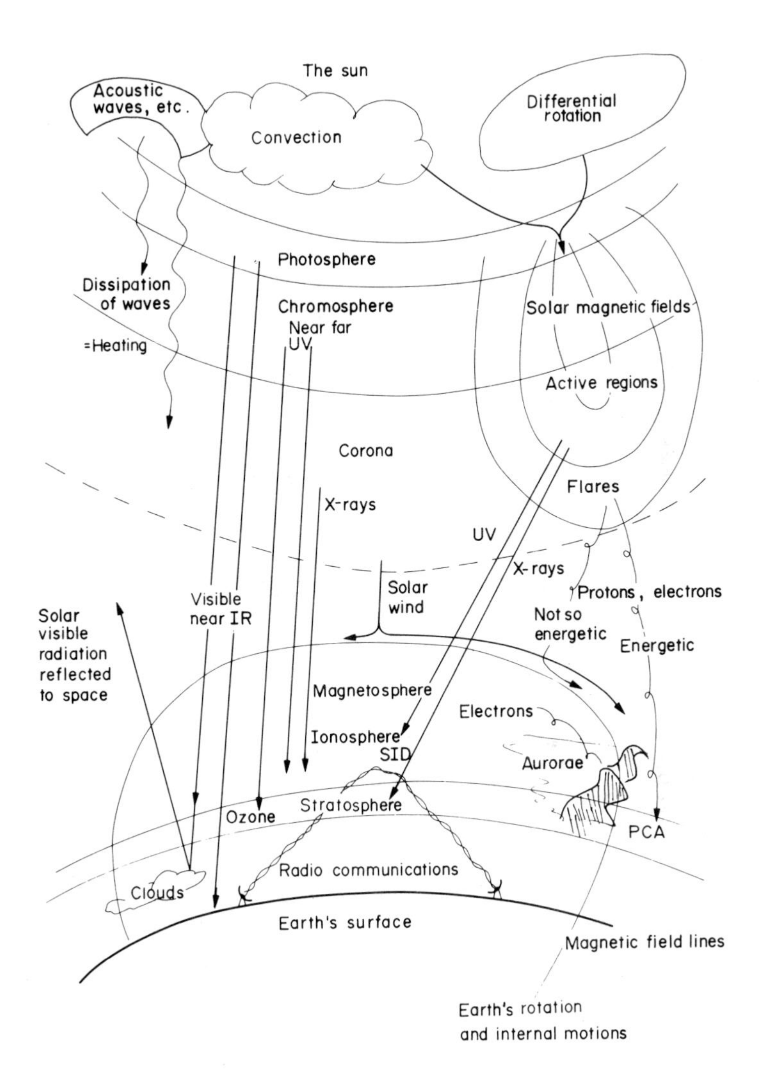

Figure 6-12. Solar-terrestrial Relations.

and heating them; the motions in this ionosphere generate various non-thermal disturbances deeper down. The visible photons, coming from deeper in the Sun and carrying most of the solar flux, also penetrate more deeply on Earth and supply the major energy input to the geosystem. Unstable magnetic field configurations on the Sun suddenly change, accelerating particles and producing a burst of X-rays and far-UV. These photons, and the particles some time later, ionize and heat the outer layers of the Earth's atmosphere and strongly perturb the Earth's magnetic field. The solar magnetic field itself, like the chromosphere and corona, seems to be a consequence of the motions in the convective zone deeper in the Sun, interacting with the Sun's rotation. Whether the motions in the "convective zone" of the Earth's atmosphere — the troposphere — can be affected by the ionosphere and magnetosphere, is at present speculative, but it cannot be excluded. Certainly the Earth's magnetic field is a consequence of the motions in the convective zone of the Earth's interior, interacting with the Earth's rotation.

If the Sun's luminosity varies, this has profound effects on the Earth's climate. Even if the Sun's total energy output remains perfectly constant, the fluctuations of its brightness at the short wavelength end of the spectrum, and the streams of energetic particles that it emits in bursts, affect our terrestrial environment. The Sun does not just heat the Earth.

7

The Energy Balance of the Atmosphere

> In tackling this theme, our starting point will be this principle:
> *Nothing can ever be created by divine power out of nothing.*
>
> Lucretius[1]
> (transl. R. E. Latham)
> *de Rerum Natura*

IN Chapter 5 we described some elements of the complexity of the climate system. We saw in particular that taking a global average, and using a single-layer approximation to represent the interaction of solar radiation with the geosystem, yields a single parameter, the global effective temperature, which does not seem to represent conditions on Earth at all well. We have already mentioned that taking into account the vertical structure of the atmosphere can give us a more informative result. The greenhouse effect leads to a surface temperature that is significantly higher than the effective temperature, the rise being limited by the "draftiness" of the greenhouse associated with convective transport of energy. In Chapter 6 we described in some detail the interactions of the solar photons with the molecules and atoms of the atmosphere and with the surface of the Earth. To work out in quantitative detail the energy flow through the atmosphere is an extremely difficult task which we shall not attempt here. Instead, let us see what we can learn from a very schematic representation or model consisting of three layers, the stratosphere, the troposphere and the surface, where we take global averages in each of these layers. Considering the "short-wave" energy flow from the Sun, for each of these layers we must ask the following questions:

How much energy is reflected upward?

How much energy is absorbed?

and for the atmospheric layers we must also ask:

How much energy is transmitted directly downward?

How much energy is scattered downward?

The numbers that I give here and in Fig. 7-1 as answers to these questions are certainly not definitive, but I believe that they are reasonably rep-

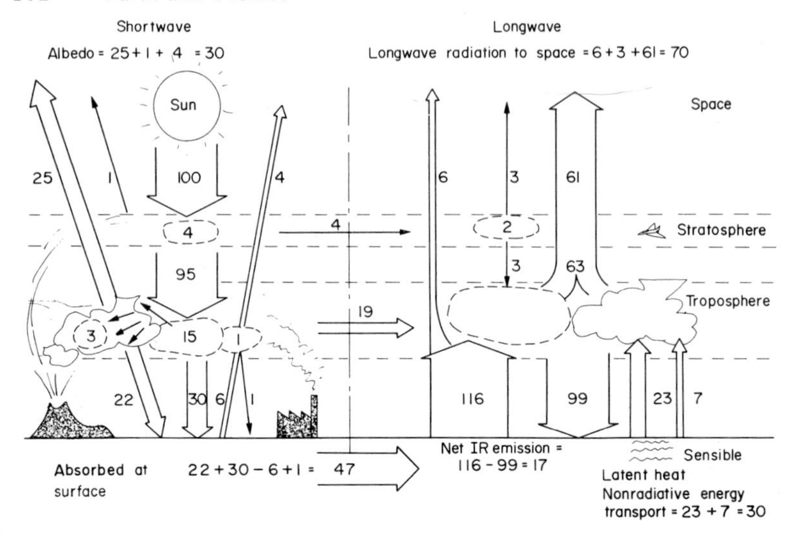

Figure 7-1. The Energy Budget of the Earth. A three-layer model.

resentative of the reality. We shall work in terms of the average solar flux outside the Earth's atmosphere. This is one fourth of what is called the solar constant, roughly equal to 1360 watts per square meter.

The first layer encountered in this model is the stratosphere. The albedo of the stratosphere depends on the amount of aerosols there, and this is known to vary as a result of volcanic activity, rising after violent eruptions such as that of Mount Agung in Indonesia in 1963. Typical values seem to be of the order of 1.5% or smaller. The stratosphere absorbs practically all solar radiation with wavelengths less that 300 nm, but very little of the visible radiation. The total absorbed then comes out to about 4% of the energy in the complete solar spectrum. Thus about 95% of the solar flux is transmitted to the layers below the stratosphere.

The albedo of the troposphere depends on reflection by clouds, Rayleigh scattering by molecules, and scattering by aerosols in suspension, many of these being of anthropogenic origin. We know that the Rayleigh scattering is stronger for the short wavelengths, giving the blue sky; clouds are relatively white. Adding up the contributions of these various tropospheric constituents, we find that about 30% of the solar flux is reflected back to space. About 19% is absorbed in the troposphere, 22% is scattered downward, while 30% is transmitted as direct (somewhat reddened) solar radiation to the surface.

Thus according to these approximate figures, only 48% of the solar flux reaches the land, ice and sea surfaces of the Earth. The average albedo of these surfaces is about 14%, but the range from one place to another, and as the seasons change, can be enormous. Working henceforth in units of 1% of the average solar flux, i.e. 1 unit = 3.40 W/m^2, this means that of the 52 units reaching the surface, 6 units are reflected back upward. The clouds and atmospheric molecules and aerosols scatter about 1 of these units back down, and they absorb about 1 unit, so that finally only 4 units of solar flux reflected by the surface reach space. A net of 47 units is absorbed at the surface.

Now we must ask what happens to the flux absorbed in the three layers of our model. Each of these layers is at temperatures well under 1000°K, so that nearly all the thermal radiation is confined to fairly long wavelengths in the infrared, quite distinct from the shorter wavelengths of the incoming solar flux. Let us see if we can balance the energy budget with this radiation. We must note that the troposphere is far from transparent at these wavelengths, principally because of the spectral absorption bands of water vapor and carbon dioxide.

Because the Earth's surface is at an average temperature of 288°K (or 15°C), it emits thermal infrared radiation at an average flux level of 390 W/m^2. Comparing this with one fourth of the solar constant we see that the surface radiates upward 1.16 times the average solar flux, or 116 of our flux units, even though only 45 units are absorbed there. Obviously there must be additional energy inputs, but they are certainly not anthropogenic.

Of the 116 units emitted by the surface, only a small fraction, corresponding to 6 units, is transmitted out to space, 110 units being absorbed by the troposphere. Thus the troposphere is nearly but not totally opaque in the infrared. The troposphere is somewhat cooler than the Earth's surface, even as seen from the ground, and sends down about 99 units of infrared radiation, which are absorbed at the surface. Thus the surface absorbs 47 units of short-wave radiation and 99 units of long-wave radiation, a total of 146 units, while it loses 116 units by its own thermal emission at long waves. This is the extent of the long-wave radiative interchange of energy between the Earth's surface and its environment. For energy balance to be maintained, as it approximately is, the surface must lose 30 units of energy flux by *non*-radiative transport.

What of the energy balance of the *troposphere*? This layer absorbs 19 units from the short-wave solar flux, and 110 units from the surface long-wave flux; it also receives 30 units from the ground by the non-radiative processes which remain to be described. It still must lose a net of 60 units of flux upward. The actual amount radiated upward is slightly greater, compensated by downward infrared radiation from the stratosphere. Let us assume that 63 units are radiated upward from the top of the troposphere. This is substantially less than is radiated downward from the bottom, but we shall see that this is reasonable for a nearly opaque layer. For the troposphere to be in equilibrium, it must be absorbing 3 units of long-wave radiation from the stratosphere.

Now we shall balance the budget of the stratosphere. Because of the low densities in this layer it is opaque only in the ultraviolet. While it is not perfectly transparent in the infrared (in particular, ozone absorbs there), it probably emits nearly as much IR flux down as up. Thus since we have assumed it to be emitting 3 units downward, it must be emitting the same amount upward, losing a total of 6 units of flux. Two thirds of this is provided by the absorption of the solar UV. The remainder, 2 units, must be absorbed from the tropospheric infrared flux. With this, our complete energy budget is balanced. The Earth emits a total of 70 units as infrared radiation to space.

We have two phenomena to explain. One is the fact that because of its opacity, the bottom of the troposphere is warmer and radiates more than the top; this is the key to the greenhouse effect. The other is the nonradiative transport of 30 units of flux, from the surface to the troposphere, the draftiness of the greenhouse.

Let us begin by examining a greenhouse with no drafts, i.e. an atmosphere in radiative equilibrium. In such an atmosphere, energy transport is by radiation alone, and the inputs and outputs must balance at every point. If say 220 W/m^2 of solar shortwave flux are absorbed at the surface and converted into heat, the same 220 W/m^2 must leave the surface and reappear as an outflow of infrared radiation from the top of the atmosphere. At the bottom of the atmosphere, this flux is a *net* flux, i.e. it is the difference between the upward and downward fluxes of infrared radiation. The temperature, on the other hand, is related to the average of the upward and downward fluxes. If for example the upward flux at a given point in the atmosphere is 620 W/m^2, radiative equilibrium requires that the downward flux be 400 W/m^2; the average value of 510 W/m^2 corresponds to a temperature of $308^\circ K$, significantly higher than the value of $250^\circ K$ which corresponds to the net flux. However, for the downward flux to be as high as 400 W/m^2, the atmosphere must be rather thick, optically speaking, since at the top of the atmosphere there is no downward infrared flux at all. The composition of the atmosphere also plays a role in determining the optical thickness of the atmosphere. In fact, the major atmospheric constituents N_2 and O_2 hardly interact at all with the infrared photons, since these molecules can spin only around one axis and vibrate only along that axis (see Fig. 2-2). It is the comparatively rare molecules made up of three or more atoms, notably H_2O, CO_2 and O_3, which make the atmosphere optically thick in the infrared and so produce the greenhouse effect, because these molecules can spin and vibrate in many different ways.

Now when we compute what the radiative equilibrium temperature structure of the Earth's atmosphere would be, we find that it would require an average surface temperature of about 330°K, i.e. 57°C. This

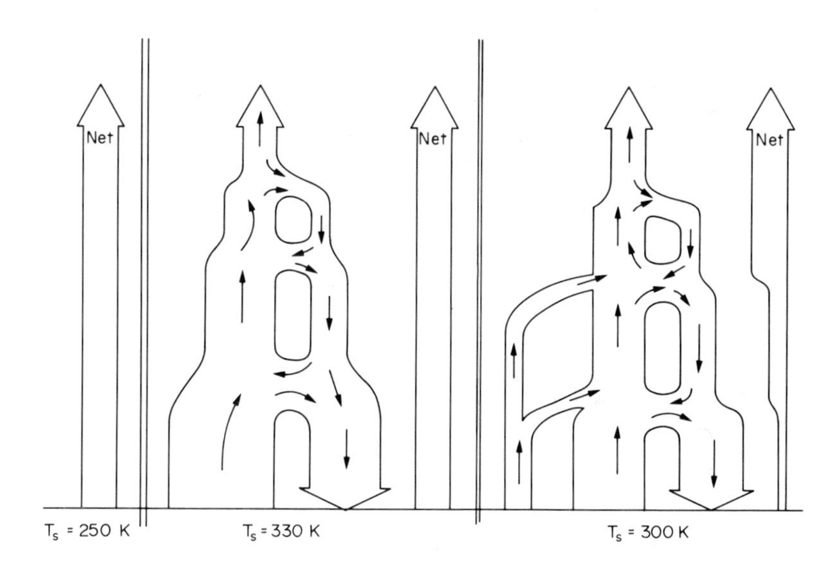

Figure 7-2. Radiative transfer of energy. Schematic energy flows from a planetary surface and atmosphere. In each case it is assumed that the planetary surface absorbs 220 watts of sunlight per square meter, and that the same net amount is radiated to space.

Left: no atmosphere — direct radiation to space. $T_s = T_{eff}$

Center: atmosphere in radiative equilibrium: the constant net flux of 220 W/m² requires T_s = 330 K.

Right: nonradiative energy transport in the lower layers allows a lower net radiative flux there, and so limits T_s to 300 K.

is much higher than the observed global average surface temperature of 15°C, and it would be extremely uncomfortable, to say the least. The Earth's atmosphere is not in radiative equilibrium. The greenhouse is drafty.

Before examining the nonradiative processes which account for this draftiness, it will be interesting to look at the energy budget of the planet Venus, which illustrates just how strong the greenhouse effect can get under appropriate circumstances. We have known for many years that the atmosphere of Venus is very thick and essentially composed of carbon dioxide. The first space probes to enter that atmosphere confirmed earlier suspicions that the surface temperature is very high, about 750°K or 477°C, much higher than the planet's effective temperature of 243°K, or even than the boiling point of water. Although obviously we still know much less about Venus than about the Earth, results from the American Pioneer-Venus and the Soviet Venera- 11 and 12 probes, all of which reached Venus in December 1978, enables us to fill in some of the details of the energy budget, as shown in Fig. 7-3.

We know that because Venus is closer to the Sun, it intercepts a solar flux that is about twice as high as that reaching the Earth. Photographs from the Earth as from close up (Fig. 7-4) show that Venus is almost

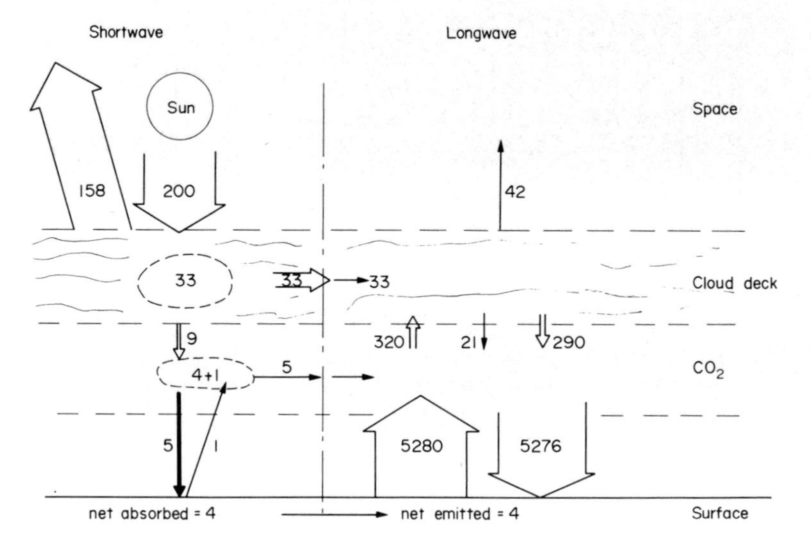

Figure 7-3. The Energy Budget of Venus. Note the factor of 20 in scales between long and shortwave.

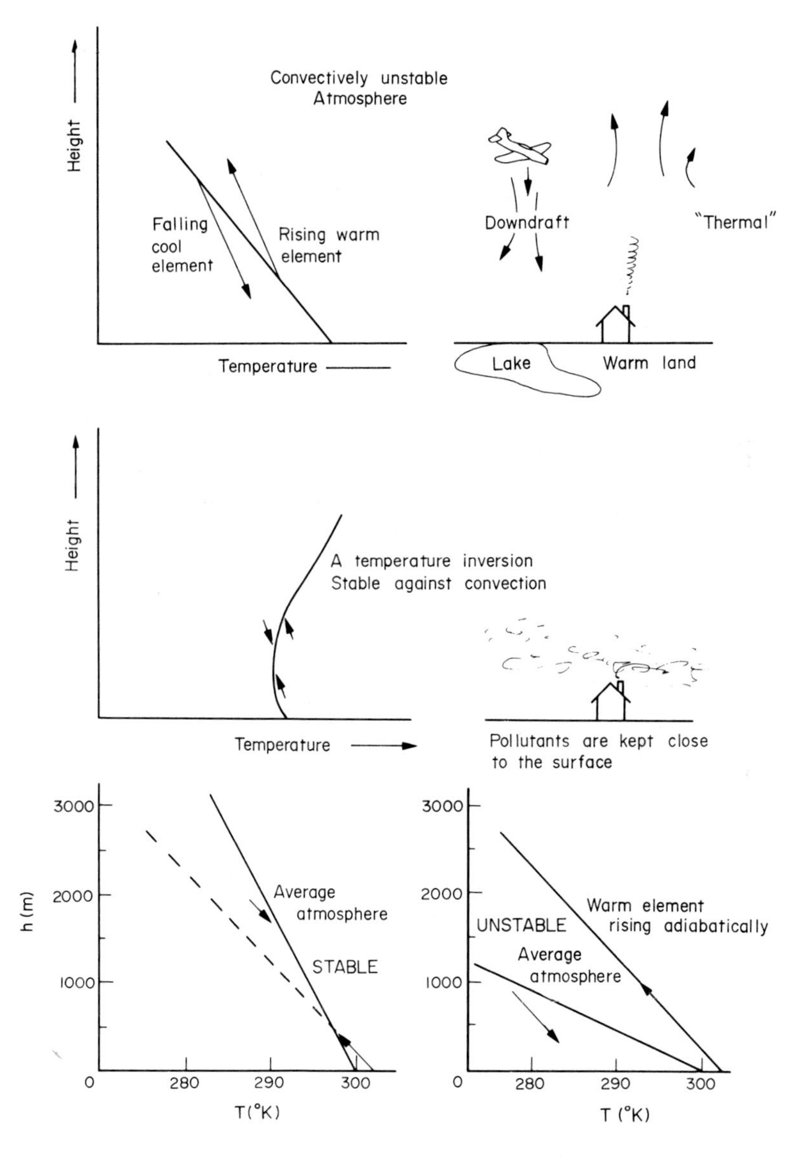

Figure 7-4. Stability or instability of a dry atmosphere.

completely covered by clouds, responsible for its high albedo. The Pioneer data[2] indicate that only 42 units (out of 200) of solar shortwave flux reach the 60-kilometer level, which is still high in the cloud layers. The clouds themselves are only partly understood, although we know that they are made principally of sulphuric acid droplets and sulphur particles, together with many other aerosols. They appear to be distributed in four different layers, each of which is extremely uniform horizontally (quite unlike the Earth), between 46 and 60 km altitude. About 33 units of solar shortwave flux are absorbed by these clouds. At 47 km, temperatures are so high that all liquid drops are vaporized, and below this level there is first a haze layer and then a clear atmosphere, with the temperature increasing steadily as the surface is approached. Only about 5 units of sunlight reach the surface, but this was enough to produce shadows seen on earlier Soviet Venera photographs. This light level corresponds to a particularly gloomy winter day in Paris or London, but in addition it is probably very reddish in cast. The surface absorbs most of this light.

Now let us consider the long-wave energy budget on Venus. With its cloud layers and its huge CO_2 content, the atmosphere is almost perfectly opaque to infrared radiation. To the surface temperature of 750°K corresponds an upward infrared flux of about 17 900 W/m^2, i.e. 5280 of our flux units. If we assume that only radiative processes take place, energy balance at the ground requires a downward flux of 5276 units from the atmosphere. On the other hand, energy balance at the base of the cloud layer, where the temperature is about 380°K, requires much lower but still substantial fluxes, about 320 units. The Pioneer data seem to show that the net radiative flux here is about 30 units, so that to balance our budget there must be a non-radiative flux downward at this level; drafts of hot air heating the lower layers. The essential point though is that because the atmosphere of Venus is roughly 100 times as massive as the Earth's, and composed mostly (98%) of CO_2, very high temperatures at the ground are consistent with a very small solar energy input there. This is the result of the runaway greenhouse.

Now let us see how the non-radiative transport of energy works. We mentioned the process of convection earlier, when we described how energy is transported upward through the outer layers of the Sun. The situation is basically the same here. In its actual state, the Earth's atmosphere is at many times and places unstable against convection. If it were to be stratified in accordance with radiative equilibrium with ground temperatures of 330°K, it would be violently unstable. Suppose that the air somewhere near the ground is somewhat hotter in one place than the average, say because of irregularities in surface properties leading to greater local absorption of solar radiation; this parcel of air will be less dense than its surroundings, it will be buoyant, and tend to rise. This may happen over a large blacktop parking lot, for example. If the rising blob of air remains less dense than its surrounding, it will continue to rise. Such *thermals* are well known to birds and glider enthusiasts. Similarly if for some reason a blob of air high up is cooler than its surroundings, it begins to sink, and as long as it remains cooler than the average at the same level, it will continue to sink. Wise pilots are wary of such downdrafts, maintaining good clearance when crossing mountain ranges. Once such convection currents are set in motion, they transport energy. The rising warm currents ultimately mix with their surroundings, and heat which came from absorbed sunlight at the surface, is finally emitted as infrared radiation high in the atmosphere. The sinking cool blobs of air may correspond to excessive radiation at high levels, the deficit being carried down to the ground and finally made up by mixing there.

Under what conditions is the atmosphere unstable? If the Earth's atmosphere were in radiative equilibrium, we have seen that the temperature would be high at the surface and decrease steeply with height, and we have said that this steep temperature *gradient* is unstable. In the atmosphere as it really is, there are places where the air is very still, and presumably the atmosphere is stable there; in other places, the atmosphere must be unstable, for convection is observed to occur. Globally, the atmospheric energy budget would not be balanced without such non-radiative transport. The important quantity is the temperature gradient or "*lapse rate*", i.e. the rate of decrease of the temperature with height (or pressure). If that is greater than a critical quantity, we have instability.

This critical quantity is called the *adiabatic lapse rate*, and it corresponds to the rate at which the temperature of an insulated blob of air drops as the blob rises in the atmosphere and adjusts to the fall in ambient pressure. The reason for this *adiabatic cooling* is that as the blob of air reaches levels of lower pressure, it expands until its internal pressure matches the surroundings. In this expansion it is doing *work* on its surroundings, and this energy can only come from the heat content of the gas. Similarly, if we compress a gas without adding heat from outside, or losing any, the temperature increases. This is why an air-conditioning system remains necessary in the pressurized cabins of high-flying jets, even though the outside air is very cold. The adiabatic lapse rate is equal to the gravitational acceleration divided by the specific heat of the atmospheric gases, and in the Earth's atmosphere it is about 10°K per kilometer.

Now we can easily illustrate the instability. We consider a point on the ground where the average temperature is 300°K, but let us assume that in a localized area, it is 302°K. The air there will begin to rise, since it will be not quite 1% less dense than the average. Let us assume that it cools adiabatically as it rises. At an altitude of 500 meters it will have cooled by 5° to 297°K, at 1000 m to 292°, at 1500 m to 287°, and at 2000 m altitude to 282°K. Suppose that the lapse rate of the average atmosphere there is only 5 degrees per kilometer. Then the average temperature falls to 297.5° at 500 m, to 295° at 1000 m, and to 290°K at 2000 m. The rising blob of air will already be cooler than its surroundings and therefore no longer buoyant, at an altitude of 500 meters. The atmosphere is *stable.* Suppose on the contrary that the lapse rate of the average atmosphere is 12 degrees per kilometer, *steeper* than the adiabatic rate. Then the average atmospheric temperature falls to 294° at 500 m, to 288° at 1000 m, 282° at 1500 m, and to 276° at 2000 m. The blob remains warmer than its surroundings, and indeed its temperature excess *increases* as it rises, making it more and more buoyant. The atmosphere is *unstable.* The same arguments can be worked out for negative temperature perturbations, i.e. for cool blobs.

All of the above is considerably simplified, and in particular it assumes that the air is dry. In reality, the air contains water vapor, the amount varying with time and place. This is extremely important, not just because we need rain for crops, but because the water cycle plays a major role in the energy cycle of the Earth. Energy is required to change the state of water, which has a very high specific heat. The processes of evaporation and condensation are particularly important. An energy of about 2.5 million joules, or 0.7 kilowatt-hours, is required to vaporize a kilogram of water, and an equivalent amount of energy is released when the water vapor condenses, say in a cloud. In the tropical oceans, enormous amounts of solar energy are thus converted into *latent heat*, transported by convection to higher levels of the atmosphere, and by the circulation of the atmosphere to higher latitudes of the globe.

This complicates somewhat the arguments relating to stability. The specific heat will depend on the water vapor content of the air, and so will the adiabatic lapse rate. Moreover, the water vapor content of the air is limited by *saturation,* which depends on the temperature. At high temperatures the air can hold more water vapor. The relative humidity often cited in weather reports is simply the ratio of the actual water content of the air to the saturation limit at the same temperature. Even a low relative humidity can correspond to a rather high water vapor content when it is hot, while the absolute moistness of hot humid air, as often encountered summers in New York, is very large indeed. On the other hand, when it is very cold, the air contains very little moisture even with the relative humidity near 100%, and if such air is heated to "room temperature", it feels extremely dry. This is common in New York apartments in the winter, and strong electrostatic charges can accumulate, leading at times to spectacular arcs when reaching to touch a doorknob. The relative humidity together with the temperature affect our comfort, particularly as the human body, like the Earth's surface, uses evaporation to transfer energy to its surroundings and so maintain an optimum body temperature. When the air is very hot, radiation is not very effective, since we absorb as much as we radiate; if it is very humid as well, evaporation following perspiration does not work either, and the combination of temperatures above 40°C and relative humidity close to 100% can be fatal if prolonged.

Since a mass of air cools as it rises, the saturation limit must drop, and the relative humidity of the air will increase until it reaches 100%. At this point, condensation to liquid water will generally take place, resulting in the formation of clouds or rain. This process accounts for the generalized formation of cumulus clouds on warm summer days in temperate latitudes, as well as for the thick cloud cover common in equatorial regions. Note on the contrary that where we have subsidence of air masses, the relative humidity decreases as the sinking air nears the ground and warms up. This is striking in the Föhn winds of the Alps or the Chinook winds of the Rockies, where the air pouring down the mountain slopes is very warm and dry.

We have not gone into any of the details of these processes, considering such phenomena as supercooling, the formation of ice and snow, the role of condensation nuclei, the different types of clouds, etc. The processes involving water are extremely important to our environment. The clouds account for most of the global albedo. The non-radiative transport (convection) of the latent heat from the surface to the atmosphere is about twice as important as the convection of sensible heat.

Water also plays a major role in the global redistribution of the incoming solar flux, as we shall see in Chapter 9. So far in this chapter, we have been using global averages, but considering vertical structure. Now let us ignore the vertical structure for a moment, and examine how the solar input depends on latitude. In the next chapter we shall see how the Earth's rotation leads to a time-dependent distribution of the solar flux over the Earth's surface. In particular we shall see how the 24-hour insolation varies with the seasons at different latitudes. We can integrate these curves over the entire year, i.e. we can add up the insolation received each day at a particular latitude. We can then plot the average solar flux (outside the atmosphere) as a function of latitude only, and we find that this ranges from 180 W/m^2 at the poles to about 430 W/m^2 at the Equator. To compute how much of this is absorbed by the geosystem, we must take into account the latitude dependence of the albedo. The albedo is rather high near the poles where the Sun's rays are usually grazing, and where much of the surface is ice-covered. It is lower at the Equator where the Sun's rays arrive more nearly vertically. The average solar flux absorbed by the geosystem, thus computed separately for each latitude, ranges from about 50 W/m^2 at the poles to slightly over 300 W/m^2 at the Equator.

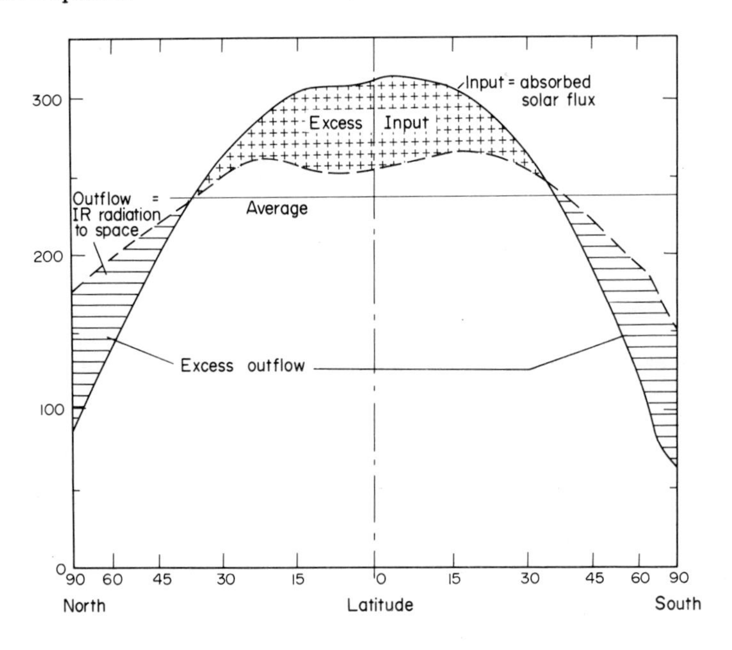

Figure 7-5. Energy gains and losses by the Earth-atmosphere system, as a function of latitude, averaged over the year (after Ellis and Vonder Haar).[3]

Using satellite measurements, we can determine the average outflow of infrared radiation from the geosystem as a function of latitude, and this has a much smaller range. About 180 W/m^2 leave the poles, and not quite 250 W/m^2 leave the Equator. The radiative input and outflow curves are strikingly different; radiative equilibrium does not hold, except on the global average over all heights. The regions between the Equator and latitudes 40° north and south absorb more flux from the Sun than they emit in the infrared. The situation is the reverse closer to the poles. Enormous amounts of energy are transported horizontally, north and south, from the equatorial and tropical regions toward the poles. In the absence of such processes, temperatures would be much higher near the Equator, and much lower near the poles, than they are now.

8

The Astronomical Rhythms: Day and Night, the Seasons, and Tides

> We cannot thoroughly understand any terrestrial phenomenon without considering what our globe is, and what part it bears in the solar system, as its situation and motions affect the conditions of everything upon it; what would become of our physical, chemical and physiological ideas, without consideration of the laws of gravitation. In the remotest case of all, that of social phenomena, it is certain that changes in the distance of the Earth from the Sun, and consequently in the duration of the year, in the obliquity of the ecliptic, etc., which in astronomy would merely modify some coefficients, would largely affect or completely destroy our social development . . .
>
> Auguste Comte[1]

IN the preceding chapters we described how the nearly constant flux of energy from the Sun determines conditions on Earth, as a result of the interaction of radiation, both solar and terrestrial, with matter at the Earth's surface or in its atmosphere. We did not consider the motions of the Earth. The solar energy intercepted by the Earth is almost constant, because neither the Sun's output nor its distance from the Earth varies very much. At a distance r from the Sun, the solar luminosity L must pass through a sphere of surface area $4\pi r^2$, so that the flux there is equal to $F = L/4\pi r^2$. Variation in the distance entails a variation in the flux according to this inverse square law. At the mean distance of the Earth from the Sun, i.e. what is called an astronomical unit (A.U.), the value of the solar flux is about 1360 W/m^2, which we call the solar constant. At the distance of Venus (0.7 A.U.) it is about twice as large, whereas near Jupiter, at about 5 A.U., the solar flux is reduced to 0.04 times the solar constant.

The orbit of the Earth is not a perfect circle, however, even though this is a fairly good first approximation both for the Earth and for the other major planets. Kepler's laws give a very good second approximation to planetary orbits. According to Kepler's first law, the orbit must be an ellipse, with the Sun in one of the foci, and not in the center. *Two* parameters define an ellipse: a size parameter, the semi-major axis, usually written a, which is half the greatest diameter of the orbit, and is the mean distance from the Sun to the planet; and a shape parameter, the eccentricity e, which measures in relative terms the distance of the focus from the center of the ellipse. Thus the maximum Sun—planet distance (aphelion) is $r_{max} = a(1+e)$, while the minimum (perihelion) distance is $r_{min} = a(1-e)$. For a circle, $e = 0$; for ellipses $0 < e < 1$. Clearly, whereas the solar flux remains constant for a planet in a circular orbit, it varies for an elliptical orbit, with maximum at perihelion and minimum at aphelion. The relative amplitude (maximum—minimum)/median of this variation can be computed from the inverse square law for flux, and is $4e/(1-e^2)^2$, which for small values of e is roughly equal to 4e.

Thus for the Earth, for which e = 0.017, the solar flux varies by nearly 7% between aphelion and perihelion, and it is only on the average equal to the solar constant. The eccentricity, and thus the modulation of the solar flux, are smaller for Venus. On the other hand, for Mars, with an orbital eccentricity of 0.093, the amplitude of the flux variation is about

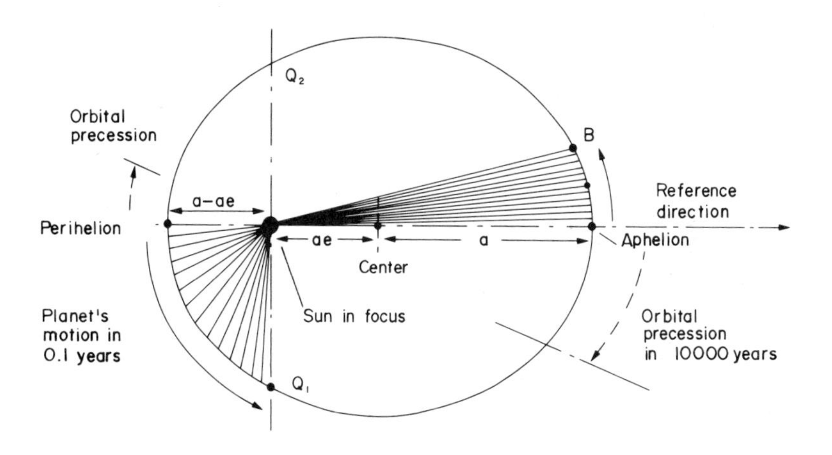

Figure 8-1. An elliptical orbit with e = 0.5. Because of Kepler's 2nd law, the planet takes the same time to move from aphelion to B as it takes from perihelion to Q_1. Planetary perturbations lead to precession of the orbit, with aphelion and perihelion moving slowly as indicated by the dashed arrows, relative to the reference direction which is fixed in an inertial frame.

37%, which is far from negligible. For Mercury, with e = 0.206, the orbital modulation of the solar flux is nearly 90% of its median value. The most dramatic cases are those of comets, with extremely elongated orbits with eccentricity nearly equal to 1. The solar flux intercepted by a comet at perihelion may be many millions of times higher than the average value it "sees" during an orbit.

In the current epoch, the Earth passes through perihelion early in January. The fact that the solar flux is 7% higher than in July may seem surprising, although it should be common knowledge that the rhythm of the seasons of the Earth is not related to our distance from the Sun; at any rate, our winter is someone else's summer. Before considering the seasonal rhythms, we might ask how fast the Earth moves in its orbit. This is given by Kepler's second law. The radius vector, i.e. the line from the Sun to the planet, sweeps out equal areas in equal times. Thus the planet moves more rapidly when it is close to the Sun, more slowly when it is far away. If we measure the angle that the radius vector makes with some reference direction fixed in an inertial frame (and remember that this links us to the whole universe), then the rate at which that angle changes is inversely proportional to the square of the Sun—planet distance. The solar flux intercepted by the planet also follows an inverse square law. Consequently, if we divide the orbit of the planet into equal orbital sectors, for example quadrants, the planet intercepts exactly the same quantity of solar energy in each of these, even though it may pass through one much more quickly than through another. A comet may spend only a few months in the two quadrants near perihelion, and 20 000 years in the other two quadrants of its orbit, but (neglecting changes in its own size) it receives just as much energy in those few months as in the remaining 20 000 years. Of course this concentrated dose of solar energy transforms the comet from a barely detectable blob in a telescope to a sometimes spectacular and beautiful object dominating the sky. For the Earth, the changes due to the orbital flux modulation are much smaller.

While Kepler's laws are a good approximation to planetary orbits, they are *exact* consequences of Newton's laws *only* when the Sun and a *single* planet, alone in the universe, are involved. Newton's law of universal gravitation states that *all* masses attract one another, and in working out the motion of the Earth, we must consider the attraction of the other planets. That Kepler's laws work so well is simply a consequence of the overwhelming dominance of the Sun, with a mass 1000 times that of Jupiter and 300 000 times that of the Earth. We can treat the gravitational effects of the other planets as mere perturbations of an otherwise Keplerian ellipse.

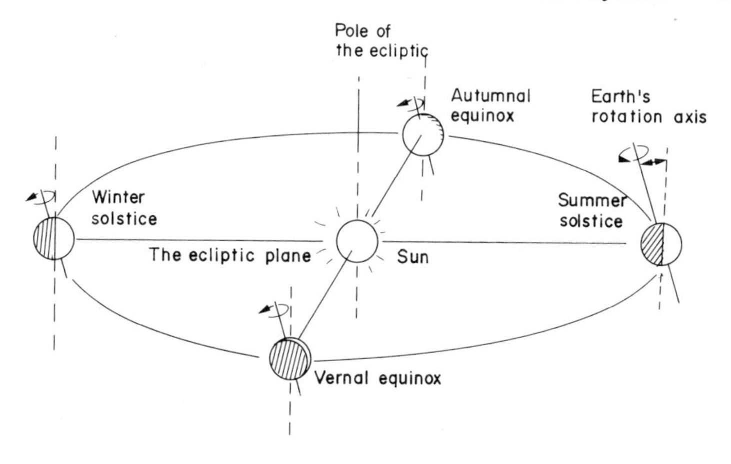

Figure 8-2. The obliquity of the ecliptic and the seasons. To a first approximation, both the Earth's rotation axis and the pole of the ecliptic maintain a fixed orientation relative to an inertial frame.

For the Earth and Mars, the major perturber is Jupiter, the most massive of the planets. The planetary perturbations lead to slight changes both in the eccentricity of the orbit and in the inclination of the orbit plane relative to what is called the *invariable plane*, which very nearly coincides with the plane of the orbit of Jupiter. The perturbations also lead to a slow *precession* of the orbit, i.e. the perihelion point advances relative to an inertial frame. Thus the perturbations produce semiregular fluctuations in the orbital parameters, with a typical period of 100 000 years for the eccentricity. These perturbations account for most, but not all, of the differences of the orbits from Keplerian ellipses. The description given so far is not complete, for we have not discussed relativistic effects. These are much smaller still than the Newtonian planetary perturbations just discussed, even for Mercury, for which they are strongest.

Before we consider the effects of these perturbations, let us look at the present state of things. The Earth moves around the Sun. We call the plane of its orbit the *ecliptic*. The Earth also rotates on an axis passing through its center, but this axis is not perpendicular to the ecliptic; it is tilted, so that the Earth's equatorial plane is inclined by about 23 1/2° relative to its orbital plane, this angle being called the *obliquity of the ecliptic*. The rotation itself takes place in about 23 hours 56 minutes when referred to an approximately inertial frame; this is the *sidereal* day, the rhythm with which the stars rise and set. It is obviously more convenient for non-astronomers, who make up most of humanity, to use a *solar* day of 24 hours, which is slightly longer than the sidereal day because of the Earth's motion around the Sun. As noted, the rotation has the consequence that stars, planets, Moon and Sun appear to move across the sky, as seen from a point fixed on the rotating Earth. If the equatorial and orbital planes of the Earth coincided (as they do for Jupiter and Venus, very nearly), the situation would be quite simple. As seen from a particular place on the Earth, the Sun would rise and set day after day, following the same path in the sky. Different stars would be visible at night as the Earth progressed in its yearly motion around the Sun, but the length of day and night would not vary, nor would the noon elevation of the Sun. Climate would still depend on latitude, the solar radiation being nonuniformly distributed over the planet, but there would be no seasons.

On Earth, as a result of the obliquity of the ecliptic, the situation is more complicated and more interesting. As the Earth goes around the Sun, its axis maintains its orientation relative to the stars, not the Sun, so that first one hemisphere, and then the other, is favored with solar flux. The apparent path of the Sun in the sky depends on the date as well as the latitude, and so the length of the day varies, as does the maximum elevation of the Sun at noon. Between the *Tropics*, i.e. the two parallels of latitude 23 1/2° north and south, corresponding to the obliquity of the ecliptic, the noonday Sun passes sometimes to the north and sometimes to the south, passing through the zenith twice a year; the length of the day does not vary significantly. Beyond the polar circles, i.e. the parallels of latitude 66 1/2° north and south, the length of the day varies enormously. The Sun does not rise at all part of the winter, nor does it set part of the summer. In the *Temperate Zones* between latitudes 23 1/2° and 66 1/2° north and south, the Sun is always seen in the equatorward direction; days are significantly longer and the Sun passes much higher in the sky in the summer than in the winter. In a table of planetary properties, the obliquity of the ecliptic may appear to be just one more number, of purely technical significance to astronomers; but change that number, and a fundamental aspect of the environment of Man on Earth is changed.

Let us examine the consequences of the obliquity of the ecliptic in more detail, first identifying points of special importance in the Earth's orbit. Shortly before the Earth passes through perihelion on January 6, it passes through the point of its orbit corresponding to the *winter solstice* (about December 21). The Earth's south polar region is then facing the Sun; as seen from the Earth, the Sun reaches its southernmost position

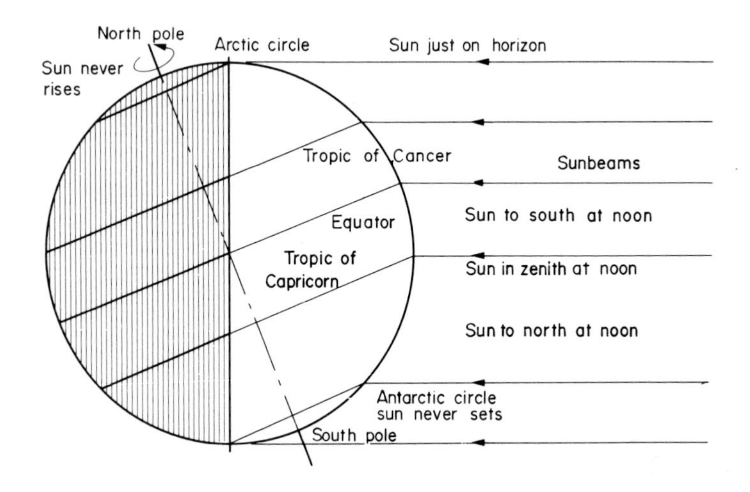

Figure 8-3. The situation at the Winter Solstice. At the summer solstice the situations north and south are reversed.

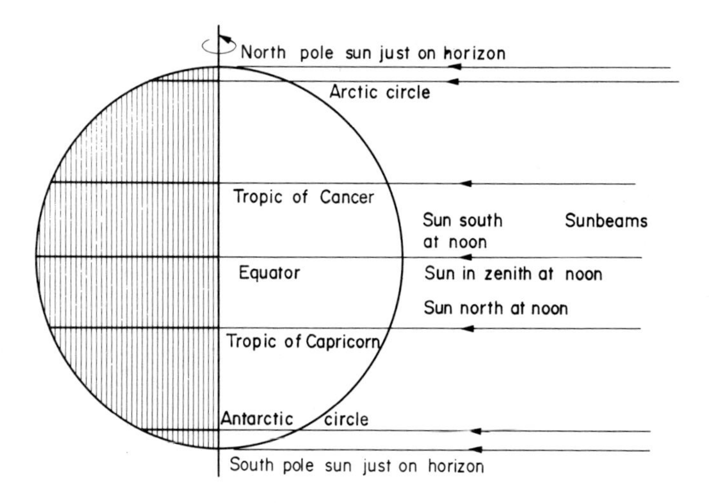

Figure 8-4. The situation at the equinoxes.

in the sky, passing through the zenith of Tropic of Capricorn, remaining above the horizon at midnight for all points south of the Antarctic Circle, not rising at all north of the Arctic Circle. It is winter in the northern hemisphere and summer in the southern. About 6 months later, at the *Summer Solstice*, when the Earth is at exactly the opposite position of its orbit, the situation is reversed, and the northern hemisphere is favored. In between, the Earth passes through the positions corresponding to the *Vernal* (about March 22) and *Autumnal* (about September 21) equinoxes, 90° from the solstices in the orbit. On these dates, the Sun passes through the zenith at the Equator, and grazes the horizon at both poles; night and day are of equal length.

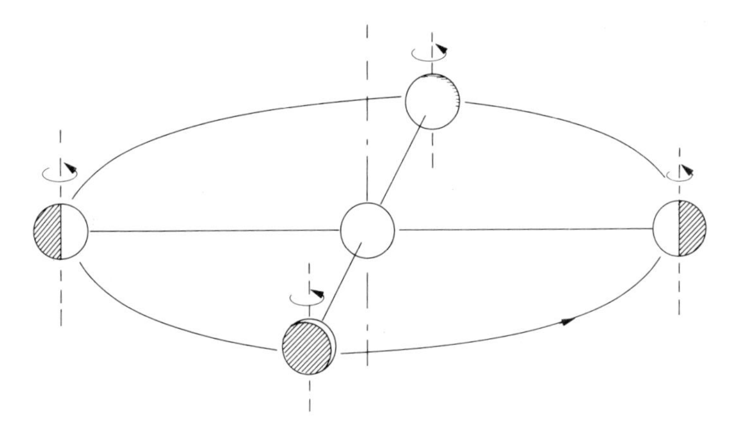

Figure 8-5. The situation for the planet Jupiter, for which the obliquity of the ecliptic is practically zero. There are no seasons, day and night remaining equal throughout the year. On the planet, the distribution of solar flux is similar to that on the Earth at the equinoxes.

If we neglect all atmospheric effects, cloudiness, etc., as well as the orbital modulation, the solar flux received by a horizontal surface is simply equal to the solar constant, multiplied by the cosine of the angle between the Sun and the zenith. For a given latitude on Earth, and a given date of the year, we can plot this instantaneous insolation as a function of the time of day. Now if we assume that this surface is continually in equilibrium with the solar flux that it receives, i.e. that it immediately radiates away in the infrared exactly what it absorbs from the visible solar flux, we can compute its temperature in the same way we computed the Earth's effective temperature in Chapter 5. For low values of the albedo (about 10% typically), we find that very high temperatures (about 390°K) should be reached at noon at equatorial latitudes, while on the other hand according to this hypothesis of radiative equilibrium the temperature should fall practically to absolute zero at night. This of course is far from true on the Earth, where there can be substantial time lags between heating and cooling, and where there are many ways of transporting energy both vertically and horizontally, not involving radiation. Maximum temperatures are often reached only late in the afternoon, minimum at sunrise after a night of cooling. At humid sea level locations, the diurnal temperature variation may be small; at high-altitude deserts it can be quite large. The situation on the Moon is much closer to the ideal instantaneous radiative equilibrium calculation described here; there is no atmosphere, and the lunar day/night period is what we call a month. Temperatures do indeed reach very high values (380°K) when the Sun is overhead; they fall steeply at night, but still they do not go below 100°K. The lunar soil has a certain heat capacity, and the heat stored during the day cannot be entirely lost even during the long lunar night. There may also be internal heat sources as on Earth.

Since the 1960's, the Moon has been part of the human domain of action, and both automatic and manned spacecraft have landed there. Thus Surveyor III made a direct measurement of the rate of cooling of the Moon's surface, when the Sun was eclipsed by the Earth on April 24, 1967 (observed from Earth as a total eclipse of the Moon); the temperature dropped from 385°K to 215°K in about three hours. The landing sites and times for the Apollo astronauts were chosen with the Sun fairly low on the lunar horizon, so as not to put undue strain on the life-support systems of the LEM and the space suits. It is amusing to note that because of the long lunar day (about 700 hours), an astronaut could keep up with the setting Sun by running, a feat that on Earth is possible only with a supersonic aircraft such as Concorde.

Returning now to our curves of the instantaneous insolation for each day of the year at a particular location, we can integrate them to evaluate the total amount of solar energy intercepted by our horizontal 1 square meter surface over a 24-hour period. We can then plot this quantity, the daily insolation at a particular latitude, obtaining a curve showing its seasonal variation. We can also show how the daily solar input is distributed over latitude. Some of the results may be surprising. Thus, in July, the 24-hour insolation does not depend very much on latitude over the entire northern hemisphere. As much (indeed more) energy is intercepted at the North Pole as near the Equator, the greater length of the summer-

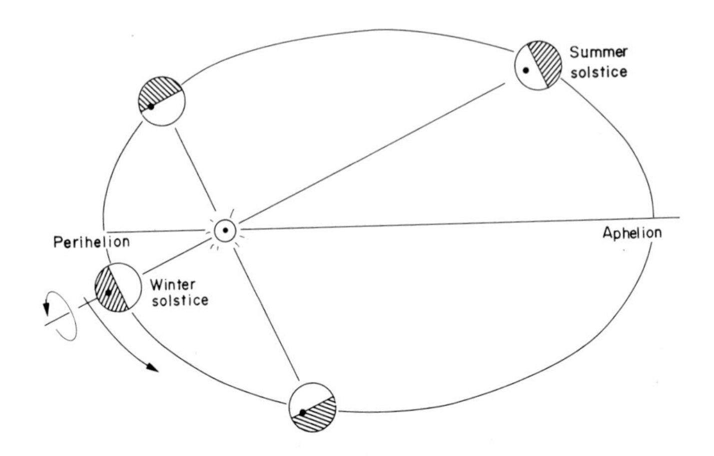

Figure 8-6. The orbit of the Earth, with the eccentricity enormously exaggerated. Note how northern hemisphere winter occurs when the Earth is close to perihelion. This was not the case 10 000 years ago.

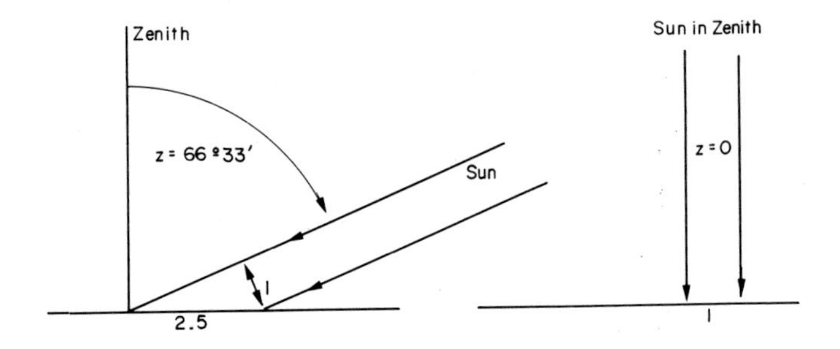

Figure 8-7. The solar flux on a horizontal surface.
Left: the situation at the Arctic Circle on the day of equinox at local noon (cf. Fig. 8-4);
Right: the situation at the Equator at the same moment.
At the Arctic Circle the solar flux is spread out over an area 2.5 times as large, i.e. $S = S_o \cos z$.

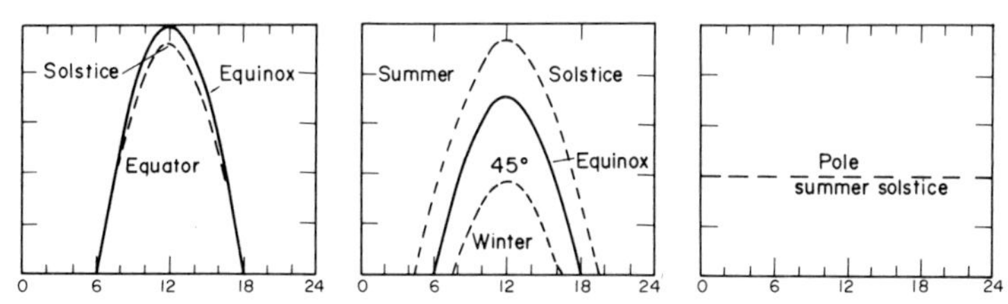

(a) The diurnal variation of insolation, for 3 different latitudes, as a function of local solar time. Full scale is 1360 W/m^2.

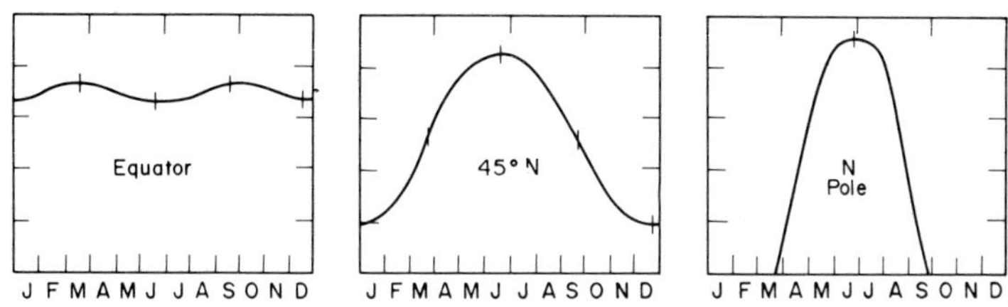

(b) The seasonal variation of 24 hour insolation, for 3 different latitudes, as a function of month. The equinox and solstice dates are marked on the curves. Full scale is 50,000 kilojoules/m^2 or an average insolation of 580 W/m^2

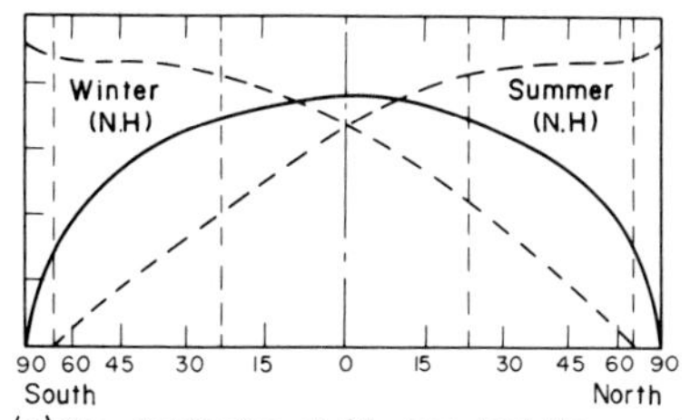

(c) The distribution of 24-hour insolation over latitude, for the equinoxes —, and the solstices --- The polar circles, the tropics and the equator are shown. Full scale as in (b)

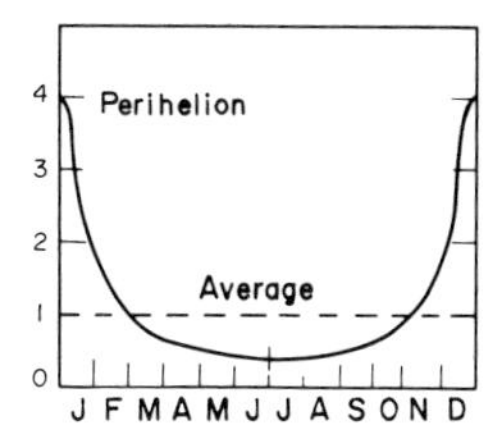

(d) The orbital flux modulation at any latitude, for an orbit with e = 0.5, and perihelion on January 1, and aphelion on July 1.

Figure 8-8. Spatial and temporal variation of the solar input (above the atmosphere).

time polar day compensating the lower elevation of the Sun there. Of course this neglects the effects of the atmosphere; the oblique solar rays near the pole suffer more absorption and scattering than the more nearly vertical rays in the tropics. At any rate, the aspect of the seasonal curves depends very much on the latitude. The insolation does not vary much near the Equator, but it undergoes an enormous variation at polar latitudes. The latitude belts, in which most of the solar flux is absorbed, are not fixed, but move north and south with the seasons, and this is of enormous importance in determining how the climate system works.

We now have examined the seasonal modulation of the solar flux intercepted as a function of latitude on the Earth, and earlier we described the orbital modulation of the solar flux received globally. The global flux maximum very nearly coincides with the northern hemisphere winter and southern hemisphere summer solstice, and if no other factors came into play, we should expect the orbital flux modulation to "soften" the seasonal rhythm in the north, and enhance it in the south. This is certainly true as far as solar flux impinging on the top of the atmosphere is concerned, but it is not at all the case for the climate observed at the surface. The fact is that at the surface, the northern and southern hemispheres are extremely different, far more of the surface being covered by water in the south.

If the Earth were relatively isolated in space, far from any other gravitating body, then its axis of rotation would retain a fixed orientation relative to an inertial frame (the distant stars and galaxies). But this assumption is absurd! Without the Sun nearby, we would not be here! Moreover, another small but not negligible mass, the Moon, is very close to us, at a distance of only 60 Earth radii. Even so, if the Earth were perfectly spherical, the proximity of these masses would not affect the orientation of the Earth's axis. However, because the Earth is rotating, it is not perfectly spherical. At the Equator in particular, the rotation produces a centrifugal force which reduces slightly the weight of the outer layers of the Earth. As a consequence, the pressure inside is reduced slightly, and the structure is modified, so that the Earth bulges slightly at the Equator. Now because of the obliquity of the ecliptic, this bulge is not located in the plane of the Earth's orbit, and there is a torque on it due to the gravitational attraction of the Sun, a force tending to "right" the Earth. Nor does the Earth's equatorial plane coincide with the plane of the Moon's orbit; the Moon exercises a torque even stronger than that of the Sun on the bulge. Because the Earth is rotating, however, the orientation of its axis is not easily changed; it reacts like a gyroscope. Just as we change our direction on a moving bicycle by leaning to one side, the effect of

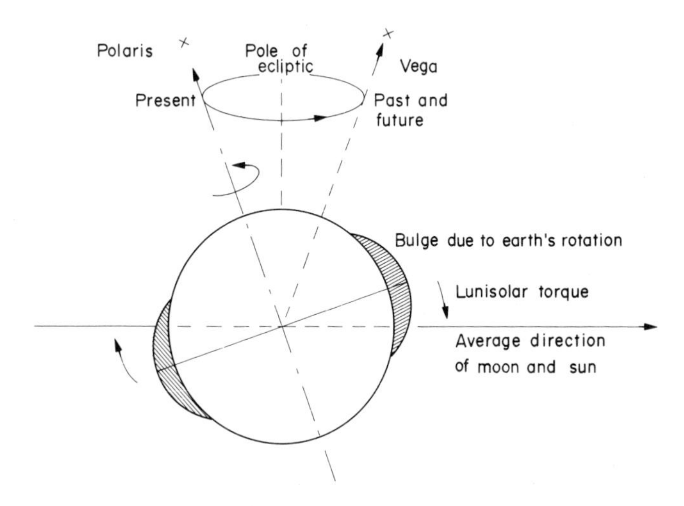

Figure 8-9. The spin axis precession of the Earth. Because of the lunisolar torque on the Earth's equatorial bulge, the Earth's axis of rotation precesses slowly about the pole of the ecliptic (roughly) with a period of 26 000 years. The pole of the ecliptic is (again roughly) fixed in an inertial frame.

the lunar and solar torques on the Earth's equatorial bulge is not to right the Earth, but rather to induce a slow conelike motion of the Earth's spin axis, analogous to what can be seen in a spinning top before it falls over. As a result of this spin axis precession, the pole star changes; the cycle takes some 26 000 years, and was already noted by Hipparchus some 2000 years ago. Thus about 14 000 years ago the bright star Vega was approximately the pole star, and not Polaris. This spin axis precession is also known as the Precession of the Equinoxes, a shifting of the position of the Sun relative to the stars on the equinox dates. Thus the Vernal Equinox point, which was located in the constellation Taurus at the time of the ancient Babylonian astronomers, has moved through the constellation Aries and is now in Pisces. As a result of combination of this effect with the orbital perihelion precession, northern hemisphere summer will take place near perihelion in about 10 000 years, and the present interaction between seasonal and orbital flux modulation will be reversed. This should reinforce the severity of the seasons in the northern hemisphere. With more summer melting, glaciers should retreat.

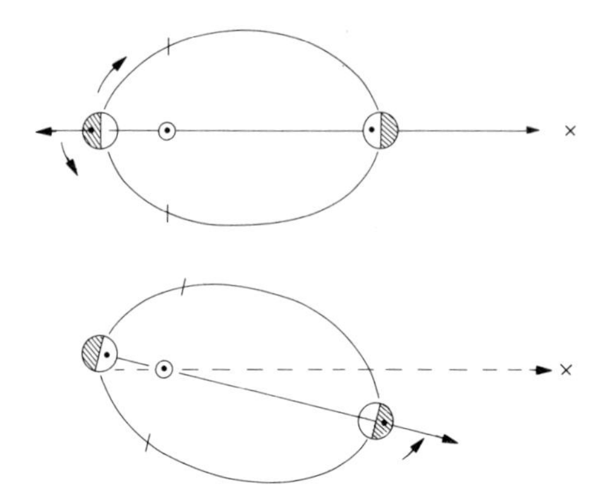

Figure 8-10. Because of the perihelion precession (due to planetary perturbations) of the Earth's orbit, taking about 100 000 years, reversal of the current situation where northern hemisphere winter solstice is close to perihelion does not take the full 26 000 years, but only about 20 000 years.

At this point it is appropriate to examine the interaction of the precession of the equinoxes with the orbital perturbations mentioned earlier. As the eccentricity e varies between 0 and 0.04, the orbital flux modulation becomes more or less severe. The obliquity of the ecliptic also varies, because of the slow oscillation of the plane of the Earth's orbit, and perturbations of the lunisolar precession. Its value fluctuates in semi-regular fashion between 22° and 24 1/2° with an average period of 40 000 years. These changes are small, but so in an absolute sense are the climatic fluctuations of the past few million years, which nevertheless have made a big difference for life in Europe, for example. Such variations of orbital and rotational parameters can be computed with confidence for the past and the future. The Yugoslav scientist Milankovitch evaluated some of the effects of these changes in work published over fifty years ago, in which he pointed out that variation in the obliquity of the ecliptic led to changes of a few per cent in the insolation at high latitudes, and that the variations correlate rather well with the glacial/interglacial fluctuations. Recent studies seem to confirm these ideas.[3] It is interesting to note

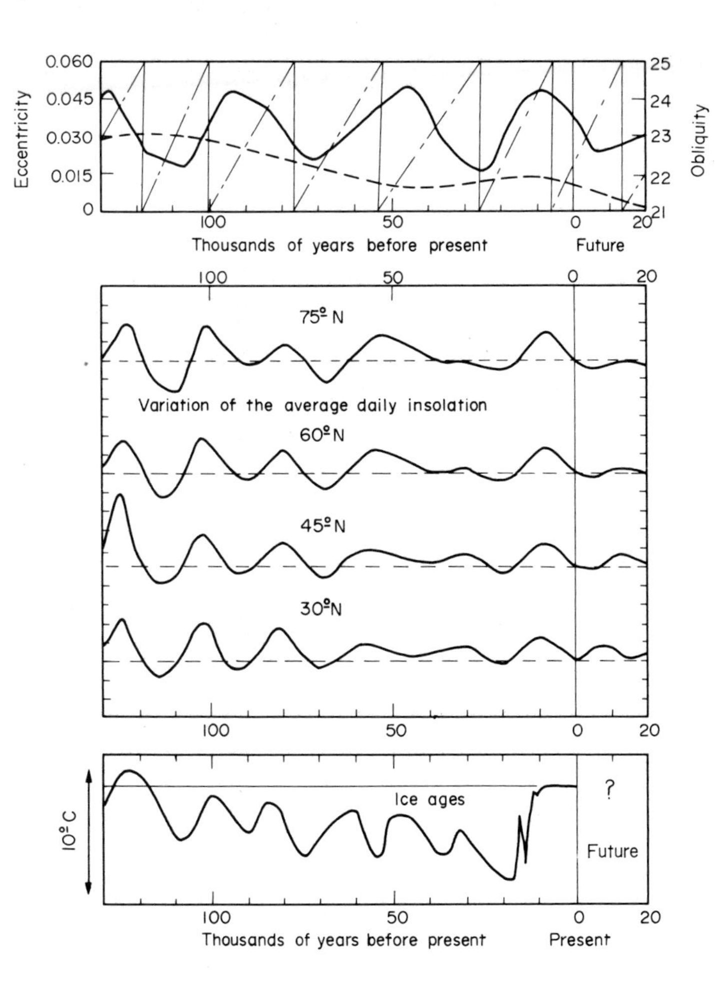

Figure 8-11. The Milankovitch theory: variations of the orbital and rotational parameters of the Earth, and climate, over the past 130 000 years, and the next 20 000.

Top: The obliquity of the ecliptic (solid line), and the eccentricity of the orbit (dashed line); the dash-dot line gives the variation of the angle between perihelion and the position at vernal equinox, presently about 90°, and going from 0 to 360° in about 20 000 years. (after Berger).

Middle: The departure of the average daily insolation from the 1950 AD values, with 1 unit of the vertical scale corresponding to 25 watts per square meter (after Berger).[2]

Bottom: An estimate of the mid-latitude air temperature in the northern hemisphere (after GARP).

that Milankovitch was able to carry on his work while a prisoner of the Austro-Hungarian Empire in World War 1, as a guest of the Hungarian Academy of Sciences in Budapest. It should be emphasized that these mechanisms may account in part for the glacial/interglacial fluctuations within an Ice Age, such as the one we have been living in for the last two or three million years, but they do not explain why this Ice Age appeared after many tens of millions of years of a much milder, completely ice-free climate. At any rate, we see that the planetary and lunar perturbations may have been a major factor in determining variations of climate over the last few million years.

This is the extent of the influence of the planets on the terrestrial environment. It is not negligible, but it remains very small, except possibly over very long time scales. There is absolutely no evidence for the major daily influence which the astrologers would attribute to the planets. This is a relic of ancient superstition and religions worshipping the planets and the stars as gods. When the Aristotelian world picture was the accepted one, there may have been some justification for a belief in astrology. In that world picture, the Prime Mover was situated beyond the sphere of the fixed stars; and all motions and happenings on Earth and in the sublunary sphere were believed to be the result of motions transmitted by the planetary spheres and the lunar sphere from that Prime Mover. Without such agitation provoked by the celestial and planetary spheres, inert terrestrial matter was not supposed to have any motion. If nature worked this way, astrology might make some sense. But we have known for at least 400 years that nature does *not* work this way, and that the Earth is not the immobile center of the universe. The laws of nature are certainly not completely known today, but our present theories of gravitational, electromagnetic and nuclear forces account far better for what is going on in the solar system than any alternative theories yet proposed, and they leave no room for the planetary influences that astrologers invoke. The gravitational influence of the obstetrician is more important than that of Jupiter! Where the gravitational influence of the planets is decisive, it is over periods of millenia, and not of days as the astrologers would have us believe.

The one aspect of astrology that might have some physical basis would involve a link to the seasons, since conditions on Earth change dramatically as the seasons change. In northern latitudes, such as Scandinavia, June marriages are traditional, and they are favorable to the extent that they lead to March births, giving the baby a period of considerable sunshine in which to grow during the first crucial six months of life. Of course today we have Vitamin D supplements. Naturally, seasonal influence over human development, if any, should depend on latitude. In any case, serious investigations of such influences should be divested of the mumbo-jumbo of the astrologers. There certainly is no reason why the advantages of spring births should have any relationship to the mythological properties of Aries the Ram that Babylonian and Egyptian astronomer-magicians saw in the stars at the Vernal Equinox several thousand years ago. In this connection it is particularly comic to watch the contortions of astrologers attempting to assimilate the precession of the equinoxes, known for the past 2000 years at least. How bitter the controversies between those who would have us stick to the Signs and those others who would consider the constellations, as though the images of the gods projected on the sky by the ancients, in a few parts of the world, had any relation to physical reality!

Coming back to reality, and to the obliquity of the ecliptic, and its impact on climate, it is particularly interesting to consider the planet Mars, whose period of rotation (24 h 37 m) and obliquity of the ecliptic (24 1/2°) are very similar to the Earth's. Mars indeed changes with the seasons, but what is observed is no longer interpreted as the growth of vegetation, but rather as sand and dust driven by seasonally shifting winds. One difference with the Earth is that the Martian orbital flux modulation is much stronger, so that the difference in climate between northern and southern hemispheres could be quite significant, especially since there are no oceans to soften them. However, there are significant differences in altitude between north and south.

The astronomical parameters for Mars also vary. The eccentricity fluctuates between 0.02 and 0.14 with a quasi-period of roughly 2 million years; thus the orbital flux modulation can be both higher and lower than it is at present ($e = 0.093$). The obliquity of the ecliptic of Mars varies over a very wide range, from a minimum of 16° to a maximum of 35°, with a period of about 160 000 years.[4] This leads to very large changes in the severity of the seasonal modulation, and this may have enormous effects on the atmosphere, whose pressure varies with the seasons. Mars also has an equatorial bulge, since it rotates. The torque exerted on it by the Sun is significant although small, while that exerted by its two tiny moons is negligible. As a result, the precession of its spin axis is very slow, taking 175 000 years. However, planetary perturbations lead to a perihelion precession taking 72 000 years, and the two effects combined lead to Martian northern hemisphere summer recurring at perihelion every 51 000 years. Generally, the fact that the Earth has a massive Moon and that Mars does not, makes a difference in the size and time scale of these fluctuations.

The theory that the climate of Mars was substantially different in the past, possibly going through cycles, is attractive in view of some of the space age discoveries concerning the Martian surface. At present, water cannot exist in liquid form on Mars, and yet many Mariner 9 satellite pictures show channels and other erosion features extremely suggestive of flowing water. Also around the polar caps there are layered terrain features that are possibly linked to successive glacial periods. Comparison of Mars with the Earth may give us further clues regarding the causes of the ice ages here.

We know very little of the surface of Venus, so comparison with this planet may be less fruitful. Both the eccentricity and the obliquity of the ecliptic are very nearly zero, so there are no significant orbital or seasonal rhythms. The rotation of Venus is more interesting: it is extremely slow, taking 243 of our days referred to an inertial frame, and its sense is retrograde (opposite to that of the Earth, or of its own orbital motion) so that a day on Venus lasts 117 of our days. Even more curious is the fact that this rotation must have been affected by the Earth through a poorly understood resonance in gravitational perturbations: Venus presents the same face to the Earth at every inferior conjunction, i.e. every time that Venus passes through (or near) the straight line between Earth and Sun.

With the exception of the tides, we have looked at the major astronomical rhythms governing the Earth, and compared our planet with others. There are smaller fluctuations which have some significance. We shall not elaborate on the lunisolar nutation, a small wobble of the spin axis as it precesses. However, let us discuss the still smaller Chandler wobble, already mentioned. If the Earth were perfectly fluid, its shape would adjust exactly to its rotation and its axis of rotation would coincide with its axis of figure. However, the Earth is solid, and in fact the two axes do not exactly coincide, the typical separation at the poles being a few meters. If the Earth were perfectly rigid, there should be a regular 10-month motion of the axis of figure relative to the axis of rotation, called the Eulerian nutation after the great Swiss mathematician Euler who predicted it in 1765. This should be detectable as slight changes in the apparent paths of the stars as observed from a fixed location on Earth, i.e. as a variation of the apparent latitude of that location. Such changes, about 0.5 arc seconds in angle, were discovered by Chandler in 1891, but they do not follow the pattern predicted by Euler. The Earth is not rigid. The Chandler wobble has two components. The annual component is related to regular seasonal motions of the atmosphere as the major zone of solar heating moves north or south. The other 14-month com-

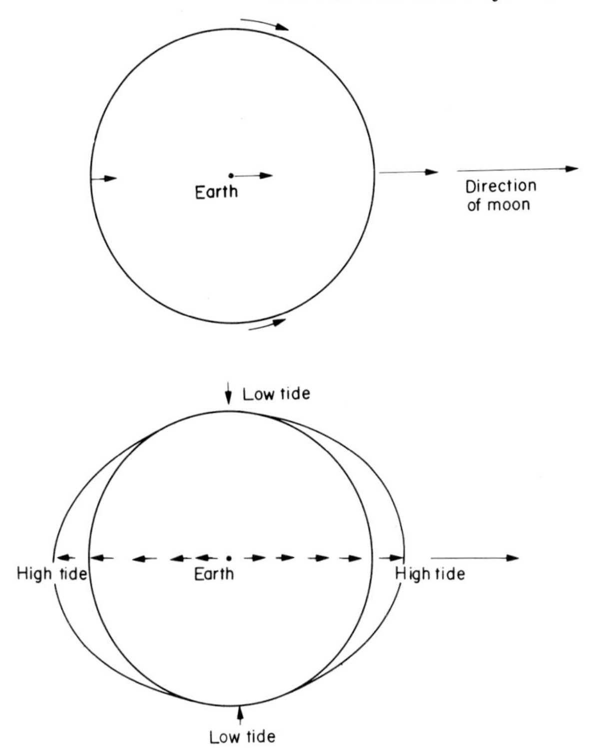

Figure 8-12. At the top, highly exaggerated, we have indicated how the direction and strength of the attractive force of the Moon varies over the Earth. At the bottom, having subtracted out the average lunar attraction, we obtain the "tide-raising force", attractive on the side facing the Moon, repulsive on the opposite side of the Earth.

ponent, alluded to earlier, seems to be maintained by internal Earth processes linked to earthquakes and volcanic activity, and by irregular atmospheric motions, i.e. the vagaries of weather.

There are also small electromagnetic forces which can affect the Earth's rotation. The solar wind exerts a torque on the tilted magnetosphere of the Earth. When particularly intense particle beams from solar flares reach the terrestrial environment, they can exert a sudden torque, giving a slight jolt to the Earth's rotation. The effect is small, but it has been

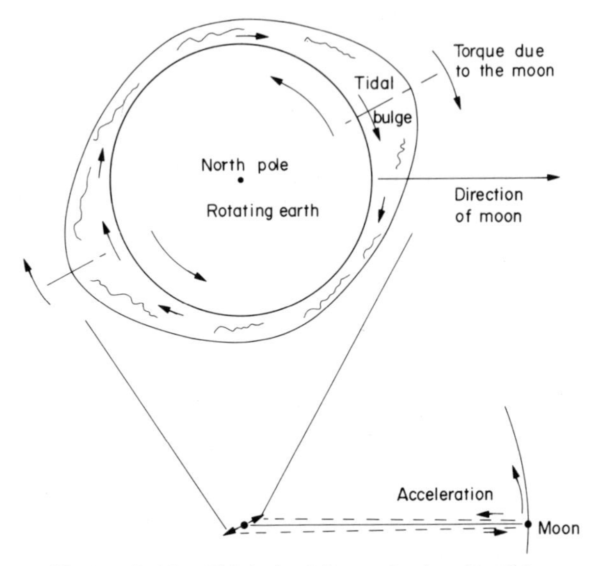

Figure 8-13. Tidal braking of the Earth's rotation, and acceleration of the Moon. The torque of the Moon on the slightly advanced tidal bulge is transmitted to the solid Earth by friction and wave action. Similarly, the net gravitational effect of the tidal bulge is to accelerate the Moon in its orbit.

measured, notably by the late André Danjon and his co-workers at the Paris Observatory. Whether such jolts can trigger earthquakes is a matter of speculation. John Gribbin and Stephen Plagemann in England have proposed a sort of scientific astrology, involving this mechanism.[5] They note that the planets, especially Jupiter, raise very small tides on the surface of the Sun, and that this may influence solar activity, in particular triggering flares. These can indeed affect the Earth. The correlations they cite between planetary tides on the Sun, and the development of solar active regions, are however hotly contested. Their "Jupiter Effect" may turn out to be bad science, but it remains a scientific approach to the problem of the influences of the planets on the Earth, and not the invocation of magic.

Now let us look at the tides. We recall that the gravitational force depends on distance, following the inverse square law. The mutual gravitational attraction between the Earth and the Moon accounts for the

Moon's motion around the Earth, or more precisely, for the motions of each of these objects around the center of gravity of the Earth-Moon system. Because the Earth is roughly 80 times as massive as the Moon, this center of gravity is very close to Earth's center, actually inside the Earth. Because both Earth and Moon are nearly spherical, for the purposes of computing their motions we can consider them to be point masses concentrated at their centers. However, we do not live at the center of the Earth. We live at the surface, and the distance of the surface from the center is by no means negligibly small compared with the Earth-Moon distance of 60 Earth radii. When the Moon is overhead, we are closer to it by 1/60th than is the center of the Earth; its gravitational pull on us is then greater by about 1/30th than the average Earth-Moon attraction. Similarly, about 12 1/4 hours later, when we are on the opposite side of the Earth from the Moon, our distance is greater by 1/60th than the center of the Earth, and the lunar pull is reduced by 1/30th compared to the average. As seen from the Earth, there is a net tide-raising force obtained by subtracting the average lunar-terrestrial attraction from the actual attraction at the point considered. This net force is one of attraction in the direction of the Moon on the side of the Earth facing the Moon, and of repulsion on the opposite side. In the fluid oceans, two tides are raised, the high tide bulges pointing (in principle) toward and away from the Moon. Because of the Earth's rotation, these bulges appear to travel around the Earth. Because of the Moon's motion around the Earth, the interval at which successive tidal bulges pass a particular point is 12 hrs 25 min, and not 12. This is well known to fishermen and visitors at seaside resorts.

The Sun also raises tides on the Earth, but its influence, only about one third that of the Moon, despite its enormous mass. This is because of its much greater distance (over 200 000 Earth radii); the relative difference in distance to the Sun for points on opposite sides of the Earth is correspondingly small. Note that the solar and lunar tides do not follow the same rhythm. Solar high tide should recur every 12 hours. The two tides reinforce each other twice a month, at New and Full Moons, when the strongest tidal amplitudes are observed, especially near Equinox. The tides are weakest when Sun and Moon are in quadrature.

All of this is an extremely simplified picture. The actual amplitudes of tides, and their times of flood and ebb, depend on many local parameters such as the configuration of the coastline. In some places, there is only one tide a day, but this is rare. Generally high tide does not occur exactly when the Moon crosses the meridian. Some locations, such as the Bay of Fundy in North America, and the Baie du Mont-Saint-Michel in France, are known for extremely large tidal amplitudes, while in the

Mediterranean, tides are barely noticeable. In the Rance estuary in Brittany, between St. Malo and Dinard, a tidal power plant has been in operation for several years, with an average output of 60 megawatts. The tides are an important environmental factor along coastlines, and harnessing them is not without environmental impact. Because of their flood and ebb, a more or less broad band of land along the shore is exposed to the action of waves. In some cases tidal currents lead to considerable transport of material, modifying the shoreline. In other cases tidal marshes provide a privileged ecological setting for a variety of living organisms.

We have already mentioned another important effect of the tides, their slowing of the Earth's axial rotation. This question was examined theoretically by the German philosopher Immanuel Kant in 1754.[6] As he noted, the tidal bulge in the oceans tends to stay aligned with the rather slowly moving Moon, but because of friction between the ocean basins and the water, the more rapidly rotating Earth must drag the tidal bulge along with it a bit. As a result, the tidal bulge is no longer exactly aligned with the Moon, but points somewhat ahead. To the Moon's attraction on the tidal bulge corresponds an equal and opposite attraction of the tidal bulge on the Moon. This should accelerate the Moon in its orbit, but the consequence of this is that the Moon moves farther away from the Earth, and ends up moving slowly in a larger orbit. If we extrapolate these tendencies backward, we come to the conclusion that in the remote past the Moon must have been much closer to a more rapidly rotating Earth, raising much higher tides. Various lines of fossil evidence tend to agree with this.[7] However, since the tidal friction must have changed enormously as sea levels rose and fell and as the continents drifted, changing the positions and shapes of the ocean basins, detailed calculation of the history of the Earth's rotation and of the Moon's orbit is practically impossible.

We have so far mentioned only ocean tides. Tides are also raised in the solid Earth. In some cases it is known that tidal strains have a triggering effect on micro-earthquakes, but it is a matter of speculation whether they are significant for major tremors. There is no dramatic correlation of worldwide seismic activity with the strength of the lunisolar tide-raising force or with its rate of change, but the idea is not unreasonable. There should also be tides in the atmosphere. Here however, both the heating effects of the Sun's radiation, following a 24-hour rhythm, and resonance effects related to the speed at which atmospheric waves travel around the Earth, complicate the picture and tend to mask the 12 1/2 hour gravitational tidal effects.

This completes our examination of the various motions and rhythms of the Earth. I have chosen to go into some detail on various points here, because it is these details which may be behind some changes in climate, and also because these details reveal the connections between at first apparently unrelated aspects of the terrestrial environment, such as earthquakes and the Earth's rotation, or the shape of the Earth and the severity of the seasons. I have not mentioned any possibilities of human impact on the motions of the Earth. There are virtually none; shifting the Earth's axis of rotation even by one meter requires energies far beyond those released by human activities, even considering the most powerful thermonuclear explosions detonated by the Soviet Union. The only possible influence of man to date would be through his triggering a major earthquake, and that has not occurred. Only minor tremors have been artificially produced to date, notably a series in the Denver Colorado area shown to be related to the disposal deep underground of nerve gas by-products, by the Rocky Mountain Arsenal. Actually there has been discussion of the possible usefulness of deliberately triggering small quakes in order to relieve stresses that might otherwise culminate in a large destructive earthquake, but in the present state of our knowledge, such attempts would be irresponsible. If we move the Earth, it is very little indeed. It is the Earth that moves us.

9

The Global Circulation of the Atmosphere

O Wild West Wind, thou
breath of Autumn's being. . .
Shelley[1]

THE weather is a changing thing, and in many ways the changes may appear to be random. And yet somehow the sum of the weather over the Earth results in a formidable organized transport of energy from equatorial latitudes poleward. Let us see how this can be. From the observational point of view, it is known that winds are not completely random. At the middle latitudes, which include New York and Paris, there is a belt of prevailing westerlies, i.e. of winds blowing from west to east. In the tropics, as described quantitatively as long ago as 1686 by the famous astronomer Edmund Halley, the trade winds blow systematically and rather regularly from the northeast or southeast, depending on one's location north or south of the equator. However, easterly or westerly flows are of no direct help in transporting energy north or south. It is the north-south, or meridional component of the general circulation that can do that.

The problem of the trade winds was examined by George Hadley in the 18th century, who related them to the greater input of solar energy at the Earth's Equator. The strong heating there leads to enhanced convection and a general rising motion of the air. This somewhat lowers the pressure at sea level. Cooler air from further north or south tends to flow toward the equatorial low pressure belt to replace the rising air. As the air above the Equator reaches higher levels, some of the solar energy stored as latent heat may be released as water vapor condenses to form clouds and rain. However, not all is lost this way, and considerable energy can be carried poleward by high-altitude flows, themselves driven by this energy release. The simplest picture of a Hadley circulation is one in which air rises over a hot surface, and sinks over a cool surface, with horizontal flow at the surface from the cool region to the hot region, going in the other direction at high altitudes. On a local scale, this sort of circulation

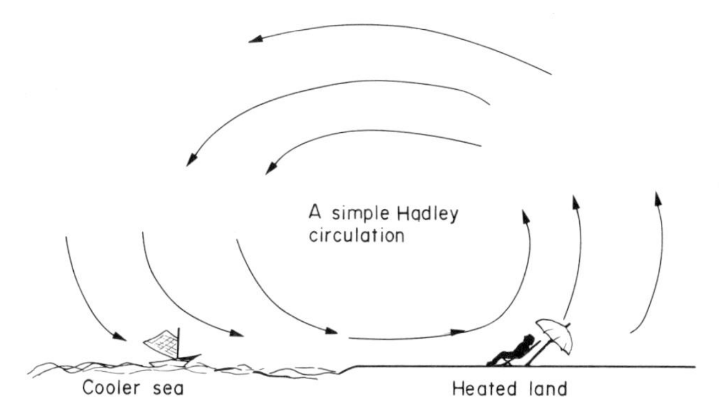

Figure 9-1. The daytime sea breeze is an example of a simple Hadley circulation, with uplift over the heated area and subsidence over the cooler area of the sea.

explains the prevalence of sea breezes at the seashore during the day, from the cooler water surface to the rapidly heated land surface.

Of course the trade winds are more easterly than they are north or south winds. This is due to the Earth's rotation. In the absence of frictional effects, the trades would be easterlies with speeds of hundreds of kilometers per hour! Hadley was well aware of the limiting role of friction; he also noted that the equatorward drift at low altitudes must be compensated by poleward drift at high altitudes, and that the frictional drag of the mostly easterly trade winds on the Earth's surface must be compensated by prevailing westerly surface winds at higher latitudes. The exact formulation of the effects of rotation on the motions of bodies on the Earth was given by the 19th century French mathematician, Coriolis. The result is that a mass moving on the rotating Earth continually experiences a force at right angles to its direction of motion, to the right in the northern hemisphere, to the left in the southern, the strength depending on its velocity as well as on the latitude. Over short distances on the rotating Earth, the Coriolis force is hardly detectable, anecdotes about it governing the direction of the swirling motion as a bathtub drains notwithstanding. When longer distances are involved, as in the motions of air masses, or of long-distance artillery, not to mention intercontinental missiles, the Coriolis force cannot be neglected. We can experiment with such forces if we work in a rapidly rotating laboratory: students might try playing a game of catch on a merry-go-round, to see the effects. Some

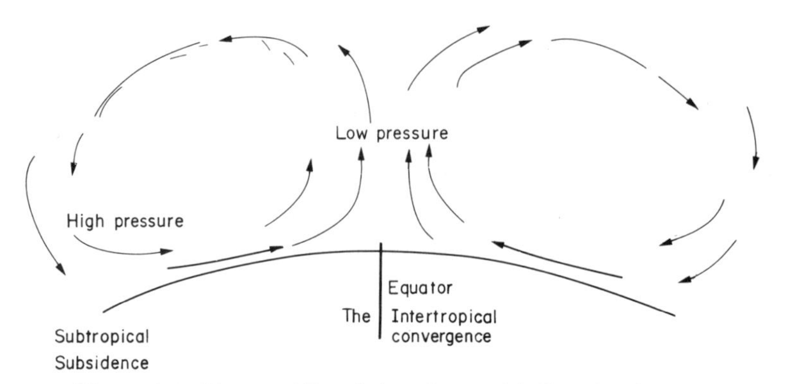

Figure 9-2. The meridional (north—south) flow in the equatorial Hadley cells.

physicists will object that the centrifugal and Coriolis forces are not real forces, that they are only fictitious forces which appear because we are working in a rotating frame and not in an inertial frame of reference. I believe this distinction to be artificial and not at all helpful. I have in this book sometimes taken the viewpoint of the distant astronomer who may well be sitting in an inertial frame. However, the atmosphere is held closely to the rotating Earth, and we should really be making things unnecessarily difficult for ourselves, were we to try to refer the atmospheric motions to an inertial frame. Let us come down to Earth; as long as we remember that it does rotate and move, we can perfectly well study the geosystem from a geocentric viewpoint.

Since Hadley's time, meteorologists have gradually developed a comprehensive picture of the average global circulation of the atmosphere. The task is difficult, because the motions of the atmosphere are unsteady. There are fluctuations on spatial scales ranging from meters (turbulence) to thousands of kilometers, i.e. the scale of the globe itself, and on temporal scales ranging from minutes to the annual seasonal rhythms, not to mention the longer scales of climate variation. Observing stations are unevenly distributed over the globe, with coverage of the oceans, particularly in the southern hemisphere, especially poor. However, considerable progress has been made in recent years, notably in obtaining data on the upper levels of the atmosphere. International research programs, such as the International Geophysical Year (IGY) in 1957—58, and the present Global Atmospheric Research Program (GARP), and UN-sponsored World Weather Watch, are providing a much better global data base, thanks to the efforts of thousands of scientists and technicians, and the coordination

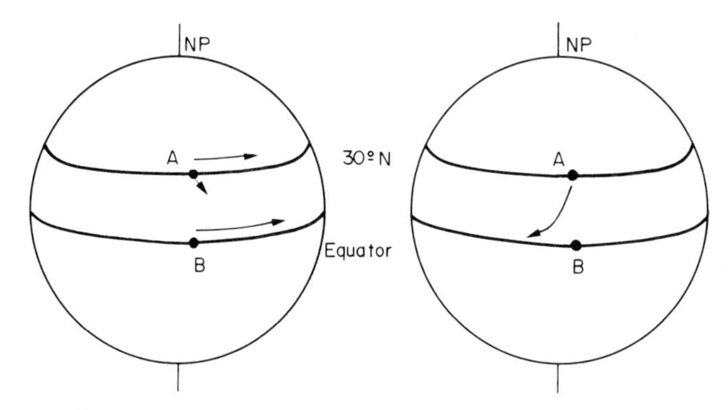

Figure 9-3. The Coriolis force on the rotating Earth.
Left: as seen in an inertial frame, a projectile fired due south from A nevertheless is moving southeast; however point B is moving still more rapidly eastward.
Right: as seen in a frame rotating with the Earth (for example from a geostationary satellite like Meteosat), the projectile starts off moving due south, but then deviates to the west.
Air masses are not projectiles, but the effects of the Coriolis force appear in their motions.

supplied by such intergovernmental bodies as the UN World Meteorological Organization (WMO) and nongovernmental organizations like the International Council of Scientific Unions (ICSU).

We obtain the average global circulation by performing averages over both time and longitude, thus considering the strength and directions of the air flows as functions of latitude and height alone. These longitudinal averages are often called zonal averages, because they apply to the entire zone situated between two parallels of latitude. Thus we make no distinction between Paris, Winnipeg or Ulan Bator, since they all lie at about the same latitude. To begin with we shall also make no distinction between the winter and summer situations, taking an average over the entire year. In this picture of the average global circulation, seasons, monsoons, hurricanes, storms and blocking anticyclones disappear. The only features that remain are the latitude-dependent climatic zones.

In the North and South Tropical Zones, we have the trade winds, which converge toward what is called the Intertropical Convergence (ITC), the belt around the Equator coinciding with the maximum average solar input. The ITC is often also called the tropical trough, since it is

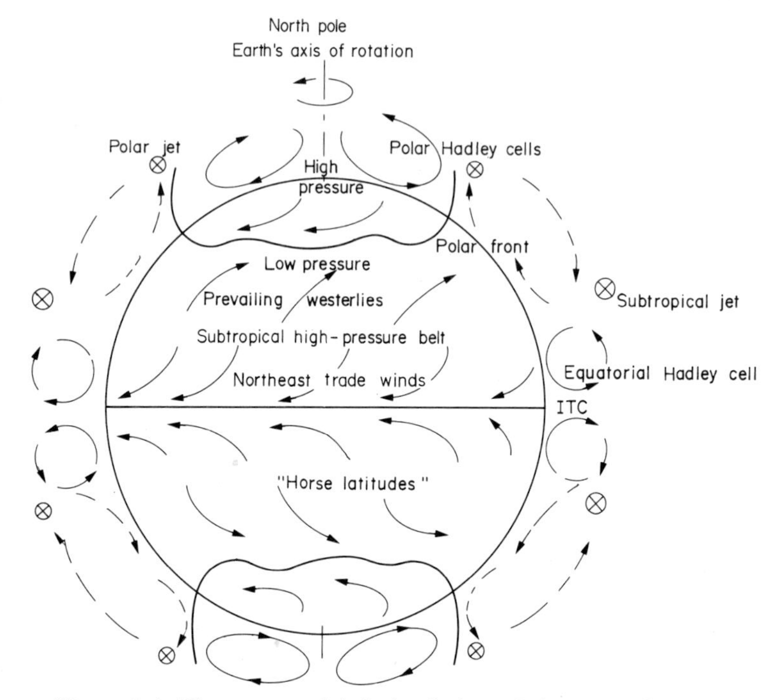

Figure 9-4. The average global circulation of the atmosphere. The Coriolis effect of the Earth's rotation is critical.

a belt of low pressure. There the convergence of the trade winds is transformed into uplift, and the heated, moisture-laden air rises. As the humid air reaches cooler levels, towering cumulonimbus clouds develop, and there are frequent violent squalls, dumping large amounts of rain in short times, releasing at the same time substantial amounts of latent heat high in the troposphere. More energy is released in a typical cumulonimbus cloud than in a hydrogen bomb explosion.

The trade winds originate in a subtropical high pressure belt, at about 30° latitude, where there is general subsidence of the atmosphere. As explained earlier, sinking air becomes relatively drier, and the subtropical high pressure belts are generally zones of deserts, all around the Earth. As the trade winds move south or north from these belts, over the tropical oceans, they pick up humidity, and so they also pick up latent heat. The poleward transport of this energy is effected in the poleward motion of the atmosphere at higher levels, which completes this "direct" Hadley circulation.

Beyond about 30° latitude, the meridional circulation is weaker, and its sense is reversed, i.e. the average drift at the surface is toward the poles rather than toward the Equator, even though the solar input continues to decrease poleward. Some of the sinking air at the subtropical high pressure belt is forced poleward, and because of the Coriolis effect prevailing westerlies develop, with speeds typically higher than the easterly component of the trades. This is because the Coriolis effect gets stronger as we approach the poles. To complete this circulation, there must be general equatorward motion at high levels in the atmosphere. This reverse cell does not seem to extend all the way to the poles. At very high latitudes, beyond the polar front, we encounter polar easterlies, part of a direct Hadley circulation with air sinking at the poles and rising at the subpolar low pressure belt.

Now this picture of the global circulation leaves out a host of weather phenomena. Furthermore, it does not account for the total poleward transport of energy actually observed in the geosystem, even including the effects of ocean currents. The tropical Hadley circulation describes the real situation up to 30° latitude reasonably well. Still, if the atmospheric motions were always equal to the average values of the global circulation pattern just described, temperatures in polar regions would still be far lower than they are now, while the tropics would be hotter. The picture of regular motions, depending only on latitude and height, satisfies neither the condition of energy balance nor those of momentum balance, even though it is a definite improvement on the radiative equilibrium picture.

The features that have been left out, the longitude-dependent migrating storms, cyclones and anticyclones, play an essential role. This is the result of instability of the global atmospheric circulation and the associated pressure distribution. Where cold and warm air masses are in contact, a perturbation of pressure in the atmosphere has a tendency to grow, forming for example a major cyclonic disturbance which transports energy poleward, both as sensible heat (warm south winds, cold north winds), and as latent heat, when poleward moving hurricanes or typhoons dump enormous quantities of rain. In the subpolar regions, particularly at intermediate altitudes, a wave-like pattern develops in the atmosphere, and instead of a subpolar low-pressure trough encircling the globe, we actually find three or four huge low-pressure regions. A variable meandering flow develops at the higher levels of the atmosphere, while down below we encounter all the vicissitudes of weather, as the big waves generate smaller eddies and finally turbulence.

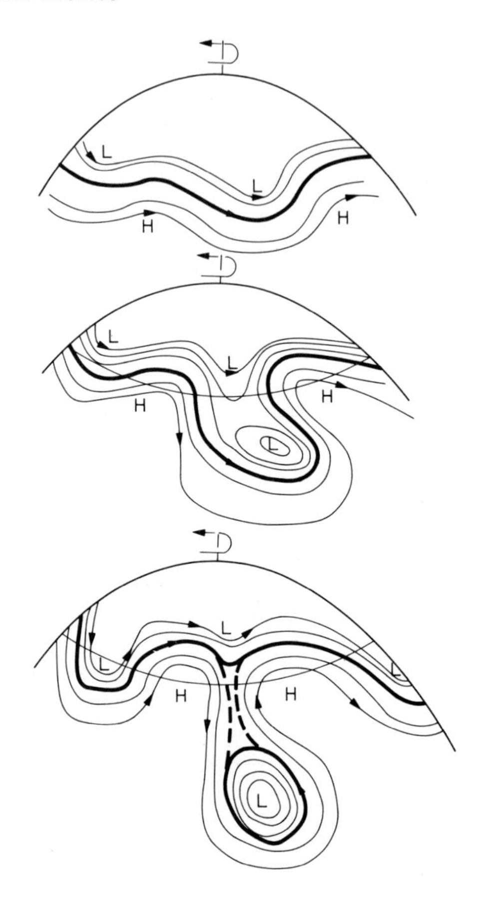

Figure 9-5. Development of a planetary-scale wave in the atmosphere (a few kilometers up, where the pressure is 500 millibars), as a result of the instability of the west-to-east flow where cold air in the north and warm air in the south meet. Note that a pool of cold air is left behind to the south, causing disturbed weather on the ground.

These processes depend on the latitude distribution of the solar input, on local irregularities, on the physical properties of moist air, and through the Coriolis effect, on the rotation of the Earth. Let us look at these effects. Suppose that over some limited area, for example in the Caribbean Sea, a low-pressure center has developed, perhaps because excess local heating leads to uplift. There will be a tendency for air to flow inward towards this center, and this air may pick up water vapor. As the air rises over the low pressure center, some of the latent heat stored in the water vapor is released by condensation, and this may reinforce the process. The disturbance can grow, the pressure in its center dropping further. Because of the Coriolis forces, the air masses being drawn toward the depression are systematically deflected to the right (since we are in the northern hemisphere). As a result, a strong roughly circular vortex develops around the depression, with strong counter-clockwise circulation of the air masses. If this disturbance develops into what is called (in America) a hurricane, the winds may be extremely strong, as high as 300 km/hr.

This is an extreme case of a cyclonic disturbance. The high winds, the heavy rains, and the rise in sea level caused by the low pressure in the storm, all combine to cause great damage when such storms reach the shore. In North America, intensive surveillance of potential and developing hurricanes, combined with effective systems for alerting the populations threatened, has succeeded in considerably reducing the loss of life due to such storms. In Asia, particularly, where large populations inhabit low-lying areas close to the sea, enormous loss of life has occurred even recently, for lack of effective means of alert and evacuation. About 300 000 people lost their lives in such a cyclone which struck Bangladesh in November 1970, and in 1977 tens of thousands were killed when a similar cyclone struck southeastern India. In the North Pacific, such storms are called typhoons, and there too they have caused great disasters. Such storms usually are born in late summer, on the western sides of the oceans, in the tropics; not only the storm winds, but also the overall motion of the storm, show the influence of the Coriolis force. These disturbances can grow as long as they are over the ocean; once over land, they are no longer effectively supplied with energy from the condensation of water vapor. While such storms may pass only infrequently in any particular place, their influence is felt over a wide area, and they are an important factor in the transport of water vapor and heat from the tropics to temperate latitudes, along the western shores of the oceans. There are of course other less violent cyclonic disturbances. Those which form farther north (south), at the polar front where cold dry air meets warm humid air originally from the tropics, are particularly important for Europe and other midlatitude west coasts. As the position of the polar front moves north and south, warm air masses may reach quite high latitudes, or cold air penetrate nearly to the tropics; in both cases we have a contribution to the net poleward transport of energy.

There also are regions of high pressure distributed over the globe, generally three giant ones along each of the subtropical high pressure belts. As already mentioned, these are regions of sinking air, of subsidence. At the surface, the air should be driven away from the high pressure center. However, because of the Coriolis force, this motion is deflected, to the right in the northern hemisphere, to the left south of the Equator. The result is an anticyclone. The winds flow around the high pressure center and not away from it, in a clockwise sense (in the northern hemisphere), roughly following the lines of equal pressure (the isobars). Although this is seldom the case close to the surface, because of friction, it tends to be the rule higher up. The pressure forces outward from the high pressure center, plus the centrifugal force which is not negligible when the winds are strong, are balanced by the Coriolis force which is directed toward the center of the anticyclone. We also can have such a balanced cyclonic wind, in which case the pressure force toward the cyclone center must be balanced by the sum of the Coriolis and centrifugal forces.

We have been discussing the role of atmospheric motions in transporting energy, in the forms of both sensible and latent heat, from place to place. However, the motions themselves must be driven, since they are limited and dissipated by friction with land and water surfaces. The kinetic energy in the motions is continually being converted into heat, and it must be replenished. We might compare the situation to that of a tanker transporting crude petroleum from the Middle East to Europe; some fuel must be used to power the engines of the tanker, but the operation is highly profitable because the energy content of the fuel needed is a very small fraction of the energy content of the cargo. The energy redistribution system constituted by the Earth's atmosphere and oceans is also quite efficient in this sense; only about 1% of the solar energy input is used to maintain the global circulation and all the smaller scale motions from cyclones to "dust-devils". On the other hand, if we define efficiency as the ratio of the kinetic energy to the total energy content of the atmosphere, i.e. if we consider the atmosphere and oceans as a heat engine, then the efficiency with which movement is generated is very low.

We have already implicitly indicated some of the places where the atmospheric motions are driven, i.e. where part of the absorbed solar energy is converted into kinetic energy. The major contribution to this

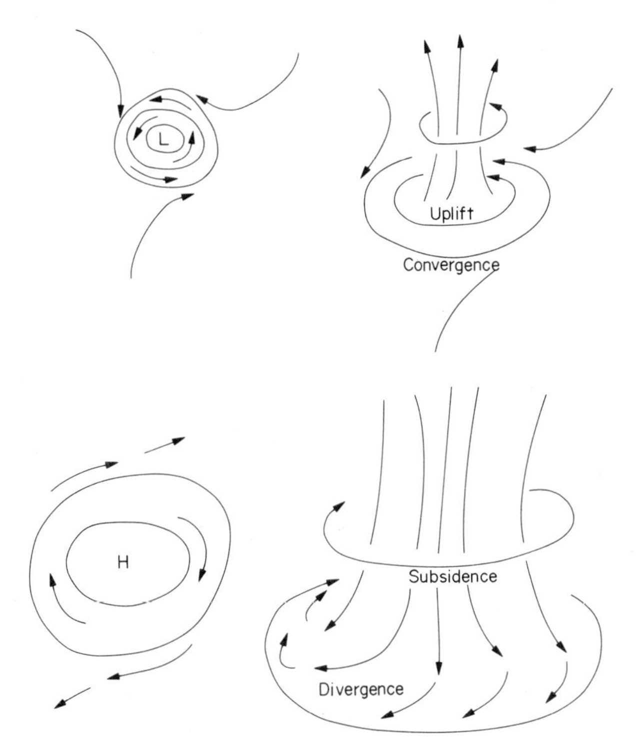

Figure 9-6. Cyclonic flow (top) and anticyclonic flow (bottom), in the northern hemisphere.

comes in the tropical Hadley cells, and more generally wherever convection is intense. In the subsiding air of a major anticyclone, gravitational potential energy is in part converted into the kinetic energy of the winds generated there, but it should be remembered that the air masses involved acquired this gravitational potential energy from solar energy absorbed elsewhere. The same is true of the water vapor which condenses and falls as rain or snow at high altitudes. Wind power, wave power and hydroelectric power are all in fact forms of solar power production.

In most of the above discussion of the global circulation of the atmosphere, we have been working with the annual averages. In reality, the uneven distribution of the solar input shifts north and south with the seasons. In a general way, the different zones of the global circulation follow this movement, although such features as the Hadley circulation weaken on the summer side. The ITC moves north and south, more or less following the latitude of maximum insolation. Seasons in the tropics are wet or dry, maximum rainfall generally occurring when the ITC passes through. Thus equatorial locations have two wet and two dry seasons, whereas at say 20° latitude north and south, there tends to be a single wet season, in the local summer, when the Sun is high in the sky.

The global circulation, by redistributing energy and water, determines what climate is like in the different parts of the Earth today. Since we know that climate has varied in the past, it is reasonable to believe that the global circulation pattern has changed too. Some attempts have been made to determine how the ice age or the climatic optimum global circulation patterns might have differed from what we observe today, but obviously such reconstructions are difficult. Similarly, there is much speculation about how the global circulation now might be changing, in connection with possible present climatic variations, such as the extended drought in the Sahel, but here we have difficulty discerning a trend in the fluctuations.

Besides transporting energy and water vapor poleward, the global circulation redistributes all other materials injected into the atmosphere: radioactive isotopes from nuclear explosions, dust from volcanoes, smoke from factories, exhaust gases and aerosols from aircraft, salt from the ocean spray, etc. Huge clouds of dust from the Sahara desert are carried across the Atlantic Ocean by the trade winds, and have been photographed from satellites as well as observed in the West Indies. The burning of high-sulphur coal in England and on the European continent, especially the Ruhr district, produces sulphates which then form sulphuric acid rain falling in Scandinavia. The dust from the Krakatoa explosion in Indonesia produced spectacular sunsets all over the world. Radioactive fallout is not the least of the hazards of nuclear war, and it would endanger nonbelligerent countries as well as targets. Understanding the circulation of the atmosphere is extremely important; we have only given its major features here, with no discussion of jet streams or of stratospheric winds in general.

Can human activities affect the global circulation? There have been more or less successful attempts at local weather modification, and these already can present complex legal or political problems, if one man's rain is another man's drought. There has been some talk of interfering with hurricanes, but the power carried by these storms makes even inveterate optimists hesitate. Of course human activities that modify the climate may affect the global circulation. There has been much discussion of one type of deliberate climate manipulation which might indeed lead to changes in the

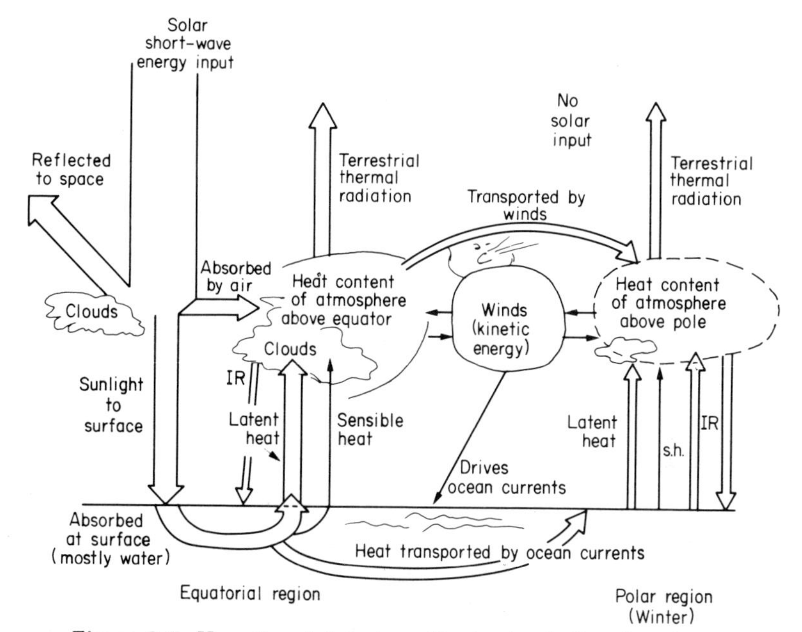

Figure 9-7. How the global atmospheric circulation functions as an energy transportation system, transferring energy from the equatorial regions where there is an abundance of solar input, to the pole, where in the winter there is none.

global circulation pattern. The floating ice pack covering the Arctic Ocean is not extremely thick, and it has been suggested that if its albedo were decreased, say by spreading black dust over it, the ice-albedo instability would lead to disappearance of the ice in a matter of only a few years. This would not affect world sea levels (unless the Greenland ice cap were also to melt), but the open ocean and the land around it — Siberia, Greenland, Canada, Alaska — might enjoy a milder climate as a result. If this were to happen — and it is by no means sure — then the gradient of temperature between Equator and pole would be gentler than now. This means that the instability which is the cause of the formation of planetary waves would be weaker. The effects on the circulation and on the generation of cyclonic disturbances are difficult to predict. Even if it were simply a matter of shifting the Earth's climate belts slightly to the north, the corresponding northward advance of deserts associated with the subtropical high pressure belt would create acute concern internationally. Deliberate manipulation of the global circulation is unlikely in the near future. It will continue to be dominated by the rotation of the planet, the amplitude of the seasonal rhythms, the solar radiation, and the internal mechanisms of the ocean-atmosphere system.

We might ask at this point how the global circulation pattern on Earth compares with that on other planets. Such a comparison is interesting because it allows us to test the theories developed to explain the terrestrial pattern, under very different circumstances. Thus on Venus, which rotates very slowly with a period of 243 terrestrial days, the Coriolis forces are very small. Close-up photographs from the U.S. Mariner 10 fly-by show an interesting banded cloud structure, and sometimes a Y-shaped feature of global scale, but virtually none of the circular cyclonic or anti-cyclonic whorls so prominent on satellite photographs of the Earth. There is evidence of strong winds at the high altitudes of the clouds, perhaps as high as 360 km/hr, and a good model of the atmospheric circulation of Venus should explain how these are driven. Measurements indicate that the poles of Venus are somewhat cooler than the equator, but they reveal very little temperature difference between the day and night sides of the planet. This should not surprise us too much, because of the enormous heat capacity of the cytherean atmosphere. It was at one point suggested that there might be a global Hadley cell on Venus. However, careful study of the cloud pictures reveals that the flow is mainly zonal, i.e. east-west. Recently theoretical calculations have shown that a global Hadley cell would be unstable, and that a pattern with strong zonal winds at high altitudes, as observed, is to be expected. These calculations also show that even the slow rotation of Venus does make a difference, and that in particular it can account for the giant Y feature sometimes seen even from Earth-based telescopes.[2]

What of the planet Mars? Here the rotation rate and thus the Coriolis forces are about the same as on Earth. Seasonal amplitudes too are similar to those on Earth. Winds do exist on Mars, revealing themselves mainly through enormous dust storms. The atmosphere is very thin, and consists almost entirely of carbon dioxide. Some latent heat is available from the sublimation of carbon dioxide ("dry ice") from the polar caps, and the release of this latent heat when the CO_2 returns to frozen form generally keeps night-time and winter temperatures from descending much below 145°K. The seasonal rhythm shows up in the atmospheric pressure measured at a ground station (the Viking lander); a substantial part of the atmosphere actually does condense at the winter polar cap. When dust is stirred up from the surface, a huge planet-wide storm sometimes develops. However, the energy source driving the wind motions in the storm is the additional absorption of solar radiation by the suspended dust, and not the latent

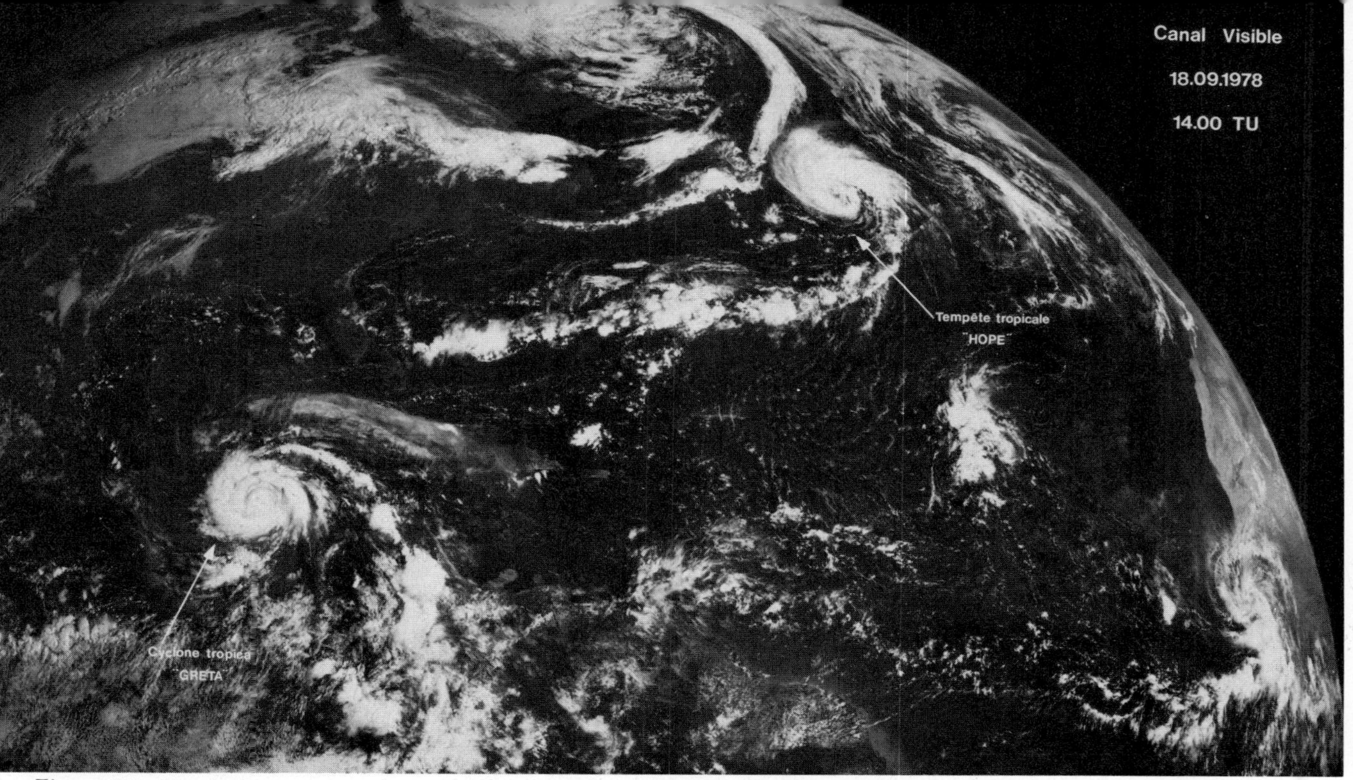

Figure 9-8. Tropical cyclones on the Earth. Hurricane "Greta" is approaching Honduras, while "Hope" is in the middle of the North Atlantic. Another storm is brewing over Dakar, to the lower right. Most of the southeast United States are clear. Image taken by the American GOES-2 satellite 36 000 km above the Equator at longitude 75° West, and received by the Space Meteorology Center in Lannion, France. Photo courtesy Météorologie Nationale, France.

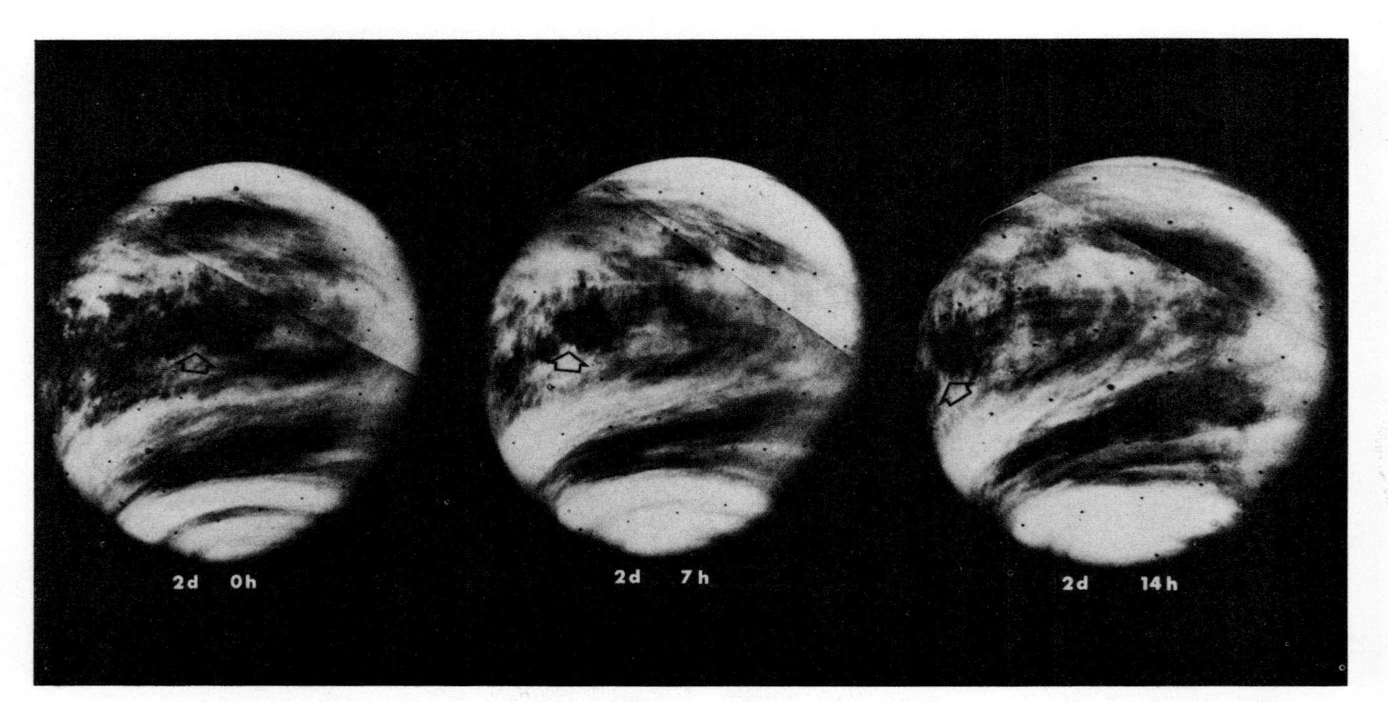

Figure 9-9. Rapid motions in the upper atmosphere of Venus. These photomosaics were constructed from pictures taken 7 hours apart on 7 February 1974, by the N.A.S.A.'s Mariner 10 space probe. Ultraviolet filters enhance contrast in the cloud deck. The feature shown is about 1000 km across. (N.A.S.A. — I.P.S. photo).

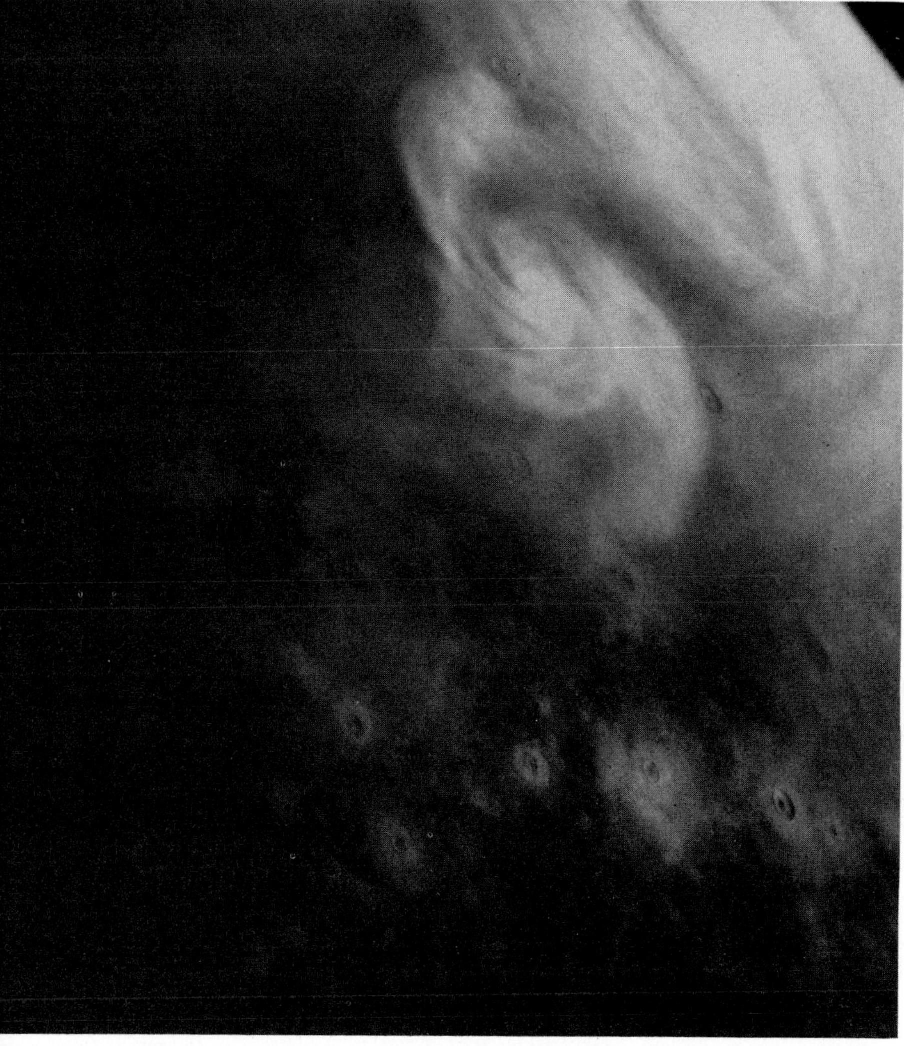

Figure 9-10. A cyclonic storm on Mars, photographed by the N.A.S.A.'s Viking Orbiter. The storm was observed during the martian northern hemisphere summer, near the north polar region, and is about 300 km in extent. North is to the top, and as it is early morning the Sun has not yet risen at the left of the photo. (N.A.S.A. and Jet Propulsion Laboratory photo, courtesy P. James).

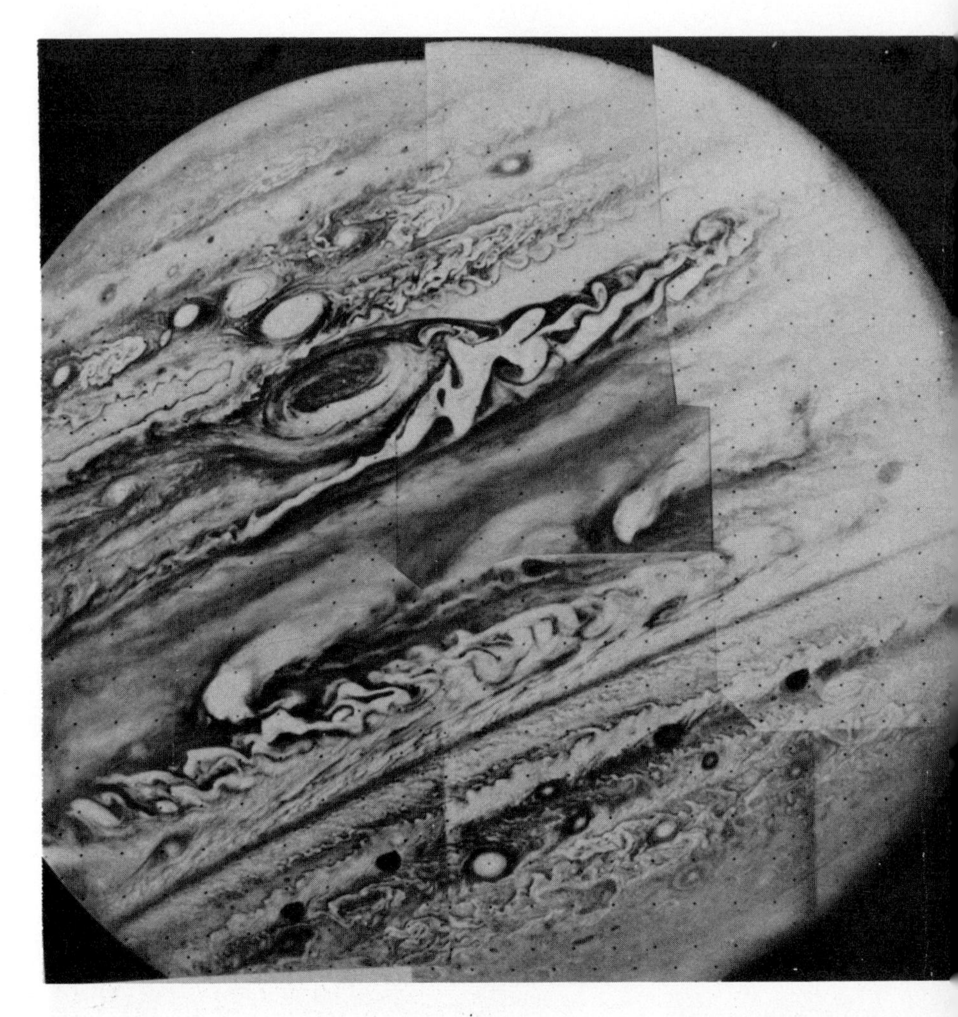

Figure 9-11. Cloud bands and structures in the atmosphere of Jupiter.[3] Mosaic assemble from photos taken by the N.A.S.A.'s Voyager 1 on 26 February 1979, with a viole filter, at a distance of 8 million kilometers from the planet. Note the extremely comple cloud formations, and the great red spot. The Coriolis effect must be very important o Jupiter, but it will be a long time before we understand jovian meteorology. (N.A.S.A — I.P.S. photo).

heat of water vapor. Theoretical weather maps can be prepared for Mars, and calculations suggest that the atmospheric circulation should depend strongly on the seasons, and also must be affected by the large altitude differences of the Martian surface. It will be interesting to test these theoretical predictions as the data accumulate. The Viking Orbiter has very recently photographed cyclonic storms quite similar to those seen on Earth, with clouds probably consisting of water ice.

10

Continents, Oceans, and Climate

> O air divine, and O swift-wingèd winds!
> Ye river fountains, and thou myriad-twinkling
> Laughter of ocean waves! O mother earth!
> And thou, O all-discerning orb o' the Sun! —
> To you, I cry to you; . . .
>
> Aeschylus[1]

UP to now in discussing climate we have mostly treated the Earth as if it were a perfectly smooth globe, surrounded by an atmosphere, but having the same properties from place to place. We know this to be false. One special feature of the Earth, which distinguishes it from all the other planets, is the existence of liquid water. The oceans cover over two thirds of the globe, but this leaves a substantial land area which contains the habitat of Man. The land areas, principally the continents, are distributed very unevenly over the globe, most of them being situated in the Northern Hemisphere, but with one continent surrounding the South Pole. The land surfaces themselves are highly irregular, with mountain chains sometimes constituting formidable barriers to atmospheric movements; half of the mass of the atmosphere is found below 6000 meters altitude, and many mountain chains rise higher than this. Even in the absence of instability, longitude-dependent disturbances in the atmospheric circulation would necessarily be produced by these surface irregularities.

The land and sea have very different properties. Water has a very high specific heat, and in the oceans this heat capacity is effectively enhanced by mixing of the layers near the surface, down to perhaps 100 meters. Considerable energy can be removed from the ocean by evaporation, and the energy removed this way is in a form that can be transported over great distances rather easily, and released elsewhere when the water vapor condenses. Moreover, the water itself can move, and ocean currents such as the Gulf Stream also transport significant energy flux poleward. By contrast, the specific heat of the land is lower and less water is available for evaporation; also, layers below the surface can only be heated by the ineffective process of conduction, and not by convection. The land can give up its energy by conduction to the air immediately above it, or by radiation to the atmosphere and to space, but neither of these processes is as effective as evaporation. Thus where strong solar flux is absorbed by a land surface, the temperature will tend to rise much more than where the same absorption occurs at the ocean surface. Diurnal and seasonal temperature variations are generally greater over land than over sea.

These effects can be seen in various aspects of the world's climates. As we mentioned in the previous chapter, the equatorial trough or ITC tends to move north or south following the zone of maximum insolation. Over the sea the amplitude of this shift is quite small: the ITC is generally very close to the Equator. Over the South American and African land masses on the other hand, the ITC dips down to latitudes 20° south in January, returning to 15° north in July. Farther east over Asia, the ITC moves far north in the summer, reaching 40° latitude over China. The distinction between continental and oceanic climates, well known, is particularly striking here. In the northern hemisphere, the lowest temperatures are not found at the North Pole, where the waters of the Arctic Ocean make their influence felt through the ice, but rather at Verkhoyansk at latitude 70° north in Siberia, where the temperature can drop as low as −68°C. The seasonal temperature amplitude is also a maximum here, reaching 100°C, the summer temperatures going above 30°C. On the oceans, by contrast, the temperature varies much less.

The relative distribution of oceans and continents determines the very large-scale seasonal variations in the circulation of the atmosphere, known as the monsoons. During the northern hemisphere winter, an enormous high-pressure region develops over the "cold pole" of Siberia. Bitter cold air spews out of this anticyclone, bringing dry cold clear weather to northeast China, Manchuria and Korea. This northwest monsoon picks up some moisture and warmth before reaching Japan, with the result that considerable snow and rain falls on the windward side of the mountains there. On the southeast coast, the weather remains sunny, as the monsoon air is dried out by its passage over the mountains. The flow then passes over the ocean, picking up moisture again. It is deflected by the Coriolis effect, and reaches south China, Vietnam, the Philippines and even southeast India, blowing from the northeast and bringing precipitation. Thus the principal rainy season in Madras occurs in winter. Of course at these latitudes, the trade winds make a stronger contribution to the northeasterly flow than does the air streaming out of Siberia.

The "true" monsoon occurs in the summer. Intense solar heating leads to the formation of a deep low-pressure center in south central

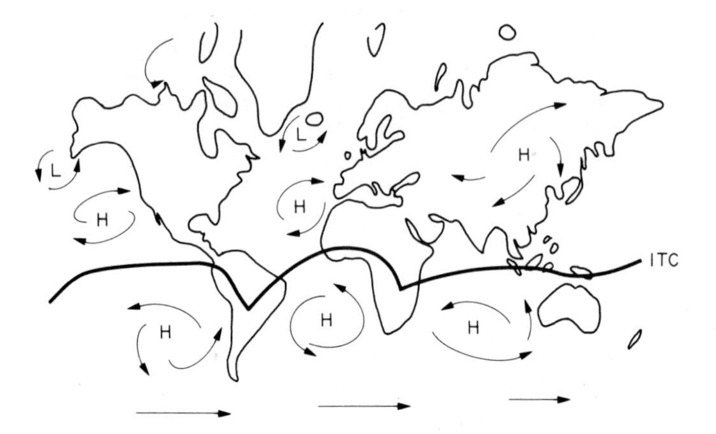

Figure 10-1. The global circulation in January. Note the shift of the ITC over the land, and the major oceanic anticyclones.

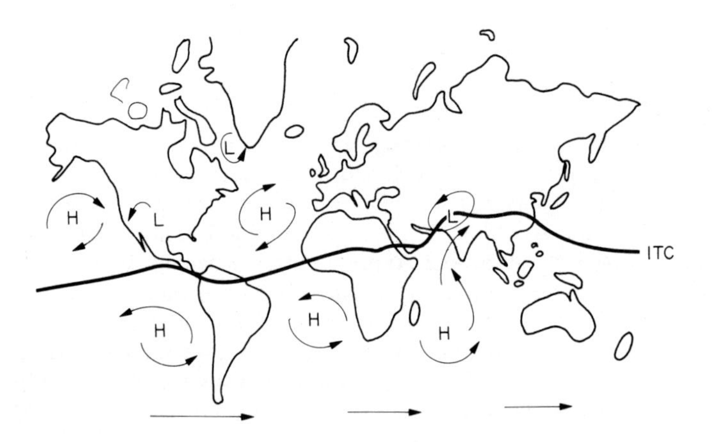

Figure 10-2. The global circulation in July. Note the shift in the ITC, and the great Indian monsoon.

Figure 10-3. Winter circulation in Asia, driven by the Siberian high-pressure center.

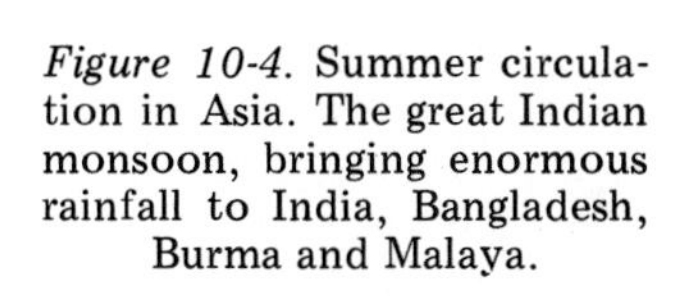

Figure 10-4. Summer circulation in Asia. The great Indian monsoon, bringing enormous rainfall to India, Bangladesh, Burma and Malaya.

Asia — Pakistan and Afghanistan especially. Air flows out from the subtropical high pressure center in the South Indian Ocean, first as a normal southeast trade wind, but it is drawn across the Equator by the central Asian low. The Coriolis force then deflects it to the right, and it reaches the western coast of India as the great southwest monsoon. The arrival of this monsoon, often a sudden dramatic event, is impatiently awaited in much of India, where it signals relief from oppressive heat and brings vital water for crops. Failure of the monsoon means enormous human suffering. The tourist visiting India with a good monsoon in August will get very wet, but he or she will see a green countryside and a happier people. Along the western coast, from Bombay down the Malabar Coast through Kerala, the humid air masses encountering the hills of the Western

Figure 10-5. Day/night changes and land/sea contrasts. These two infrared views from Meteosat show the hottest areas as darkest.
Left: early afternoon. The Sahara is much hotter than the surrounding ocean.
Right: midnight. The Sahara, Egypt and Arabia are now significantly cooler than the sea. Note the changes in cloud patterns in the 9-hour interval between the pictures.

Note also the rather well-developed ITC over the Atlantic, showing up as a narrow east—west band of cloud. Clouds cover much of western Europe, but where it is clear, Scotland, Denmark and southern Sweden appear lighter, i.e. colder, than the surrounding sea; this is reasonable for the date of the images, i.e. February 19. (Meteosat photos courtesy European Space Agency).

Ghats dump enormous amounts of rain. Behind these hills, particularly on the South Deccan plateau, and along the southeast coast from Madras to Pondichéry, conditions are much drier, without being arid. The southwest monsoon waters nearly all of India. Where the airflow encounters the colossal Himalayan barrier, as in Assam, rainfall reaches record levels; in July 1861, 9.3 meters of rain fell in Cherrapunji. The summer monsoon of India is the most dramatic example of the influence of land and sea on the global circulation, acting in concert with the seasonal rhythm. For several months, the "normal" pattern is reversed, the effects even being felt in the currents of the Arabian Sea.

Less dramatic but somewhat similar changes occur elsewhere. The Sahel droughts are in fact failures of the West African monsoon. A low pressure center develops in the summer over the southwest United States, particularly Arizona, and what little precipitation occurs there usually falls in August. During this month, observing conditions for astronomers at the Kitt Peak National Observatory are not very good. Nothing really comparable to the great Indian monsoon develops, however, in part because the high mountains of the coastal range rather effectively isolate this region from the Pacific.

In the winter, strong high pressure centers often develop over the northern United States and Canada, and the resulting anticyclonic circulation brings cold sparkling clear dry weather to the northeast United States, much in the same way the northwest monsoon from Siberia affects northeast China. In Boston, the cold northwest wind is sometimes called the "Montreal express". There really is no mystery about why New York, at the same latitude as Figueira da Foz, on the eastern shore of the Atlantic in Portugal, is so much colder in the winter. This is simply determined by the distribution of land and sea, and by the fact that eastern and western shores are different because of the Earth's rotation.

The distribution of land and sea interacts with the latitudinal distribution of solar input, and with the Earth's rotation, through the global circulation pattern, in a way which accounts quite well for the distribution of precipitation over the Earth. We have mentioned that deserts tend to lie along the subtropical high pressure belt, where air is generally subsiding and drying out. This is striking on the map of Africa and Arabia. However, further east we see the effects of the monsoon watering India, and both monsoons and typhoons watering the subtropical regions of China and southeast Asia, situated as they are on the western shores of the Pacific. The desert belt is located further north in central Asia.

In North America, desert appears on the southern end of Baja California, at the Tropic of Cancer, but otherwise the only deserts are located farther

Figure 10-6. The major world deserts. Major mountain barriers are indicated.

north, in the Great Basin between the Sierra Nevada and Rocky Mountain chains. Here the north-south orientation of the major mountain ranges limits the penetration of moisture from the Pacific. The east-west orientation of the Himalayas plays a similar and more pronounced role in preventing Indian Ocean moisture from reaching Tibet, but it also shields northern India from the direct effects of Siberian air in the winter.

In the southern hemisphere, deserts also appear along the subtropical belt, but only on the western side of the continents. The extreme is reached in the Atacama desert of northern Chile, where less than a millimeter of rain falls in a year, and many localities have no rain at all for a decade or longer. Several major international observatories have been established in recent years at the southern limits of this region. Southwest Africa and much of central and western Australia are also deserts. The eastern parts of these continents are not deserts even at subtropical latitudes, receiving some moisture from the trade winds. There is no continuous high pressure belt, but rather a more or less permanent anticyclone over each of the three southern oceans. Most of the southern hemisphere is water, but the tip of South America does extend to the mid-latitude regions, where prevailing westerlies blow. There, the western coast (southern Chile) is well watered, while the eastern part — the Patagonian desert of southern Argentina — is extremely dry. The contrast is enhanced as a result of the barrier constituted by the southern Andes mountains.

The mid-latitudes of the northern hemisphere include most of North

America, practically all of Europe, and Siberia and much of China in Asia. Deserts are found as mentioned before in central Asia, shielded from humidity by the great extent of land east and west, and by the high east-west mountain chains to the south. The western shores of the continents are well watered by more or less modest cyclonic disturbances picking up water from the ocean and carried along by the prevailing westerly flow. In northern California, Oregon, Washington state and the Canadian province of British Columbia, as in northwest Spain, the predominantly north-south relief causes high precipitation close to the coast, with regions farther east relatively dry. In Europe, central Spain is in a rain shadow, but elsewhere moisture can reach points far inland. There is considerable precipitation in the Alps, but since their orientation is mainly east-west, they do not present a major obstacle, although their shielding effect is felt on the Mediterranean shores to the south. The climates of mid-latitude western shores are generally dominated by the oceans. In western Europe the cyclonic disturbances spawned in the winter by the subpolar Icelandic low make weather very difficult to predict, although the data obtained from satellites is far more complete than what was obtained from weather ships, balloons and airplanes in the recent past. In the summer the tracks of such storms depend on the position of the semi-permanent high pressure center over the Atlantic. When this Azores anticyclone migrates to the northeast and settles down over northern Europe or England, long periods of warm dry sunny weather result, as in the remarkable drought of 1976.

Since the prevailing flow in the mid-latitudes is westerly, it is easy to understand why the climates of the eastern coasts of continents are so extreme, with hot summers and cold winters. New York is not to be compared with Bordeaux in France on the eastern shore of the Atlantic, but rather with Tientsin in China on the western shore of the Pacific. It is not so evident why these eastern coasts are as well watered as they are, if we consider only the average westerly flow. This again illustrates the inadequacies of averages; most of the rainfall comes from the penetration of masses of moist air from the subtropical oceans, and it falls in concentrated amounts. Indeed a substantial fraction is dumped by tropical cyclonic disturbances (typhoons, hurricanes) which reach the western shores of the oceans somewhere in the mid-latitudes. These influences can extend far inland, in the absence of high mountain barriers. Thus the Great Plains of the central and western United States, shielded from the Pacific by the Rocky Mountains, depend on water from the Atlantic and the Gulf of Mexico. In the 1930's, such penetration failed, leading to prolonged drought and the formation of the dust bowl.

We have seen how the uneven distribution of land and sea, of mountains and plains, modifies the climate pattern that we should expect if we considered only astronomical parameters and atmospheric physics, as in the preceding chapter. Let us recall that all these features are determined by the internal processes of the solid Earth, responsible for the drift of the continents and the uplift of mountains. We have not yet looked in more detail at the oceans, although we have pointed out that much of the incoming solar energy is absorbed there.

When we evaluate the energy actually carried poleward by atmospheric motions, we find that it is not sufficient to make up the difference between solar energy absorbed and long-wave radiation emitted. Ocean currents must carry a significant amount of energy away from the tropical zone, especially at latitudes about 20°. Cold and warm ocean currents are of course known to exist in surface layers. The best known is the Gulf Stream, beginning off the southern end of Florida, moving north along the east coast of the United States and then veering northeast across the Atlantic, ultimately warming Brittany, the British Isles, Iceland and Norway. In fact the Gulf Stream is part of a vast clockwise circulation around the North Atlantic, other parts being the Canary current moving south along the west coast of Africa, and the North Equatorial current moving west across the Atlantic. A somewhat similar clockwise gyre dominates the North Pacific, while counterclockwise gyres fill the South Pacific, South Atlantic, and Indian Oceans.

The sea-surface circulation pattern thus conforms to the Coriolis effect. In fact it is determined by the driving action of the prevailing winds together with the barriers of the continental shores. Thus below the latitude of Cape Horn, where the southern ocean is uninterrupted by land, the prevailing westerlies drive what is known as the West Wind Drift. At the Equator, the converging northeast and southeast trade winds drive the equatorial currents moving from east to west. When these currents encounter the continents, they obviously must bifurcate, or reverse to form a counter-current. At any rate, warm water is driven poleward, either north or south. In the North Atlantic this forms the Gulf Stream, with a sharp limit between warm water from the south and cold water from the north, easily detectable on infrared images from satellites, and also manifest at the surface by persistent fog banks where warm humid air from the Gulf Stream is cooled to saturation by the cold water to the north. In the North Pacific, the corresponding current is the Kuroshio, which warms the Aleutian Island chain but is prevented by them from reaching as far north as does the Gulf Stream in the Atlantic. These warm currents provide energy to the air masses above them, both directly as sensible heat and in the form of enhanced evaporation. At the eastern shores of the oceans, these humid air masses and storms regularly provide warmth and water for the land.

When these currents turn south (or north in the southern hemisphere), they have lost some of their warmth. Depending on the configuration of the coastline, on winds, and in accordance with the Coriolis force, these currents may be driven away from the land, and this drags up cold deep ocean water in what is known as upwelling. Thus the south-flowing California current and Canary current are cold currents. The upwelling brings up sediments accumulated on the ocean floor, which contain nutrients essential to life. The result is extremely rich marine life in these regions, at all levels of the food chain from microplankton to fish to seagulls, seals, and humans. The most dramatic example of this is in the north-flowing Humboldt cold current off the coast of Peru, where upwelling provides nutrients for the world's largest anchovy fishery, a major source of protein. Despite recent progress in both theory and in satellite observation, the variability of this upwelling is not yet fully understood. Every four to seven years, a phenomenon occurs which is called El Niño (the Christ child) because it first appears around Christmas. Warm water invades the region, the winds and currents change, upwelling fails, and the anchovy population drops catastrophically. In the past, this was an accepted aspect of the economics of the anchovy fishery, and caused only minor difficulty since the anchovy population recovered each time. Recently, however, perhaps because of overfishing, the anchovy population has not recovered, leading to severe economic distress. The details of ocean currents are important and complicated, involving small-scale eddy motions, and the picture of the smoothly circulating giant gyres is as unrealistic as the picture of a global atmospheric circulation without anticyclones or storms.[2]

There is circulation at greater depths as well, the densest water sinking to the bottom. The density of sea water depends on both its temperature and its salt content, and generally the coldest, saltiest water is densest. Now when salt water freezes, very little of the salt goes into the ice, most of it increasing the salinity of the remaining water. Thus at the edge of the Antarctic ice shelf, very cold extremely salty water is continually being formed, sinking to the sea bottom. Of course it mixes on the way down, and sediments at the bottom are stirred up making Antarctic waters extremely rich in life. Nevertheless, a distinct layer of Antarctic Bottom Water is formed, which spreads out northward along the ocean bottom and indeed has been identified as far north as Hawaii. These deep-water currents are very slow, and circulation times of hundreds of years are involved. This is one of the reasons why it is so difficult to

Figure 10-7. Principal ocean currents.

calculate the evolution of climate, since the present state of the oceans, which is a major determinant of the present climate, and which will influence future climate, depends on the climates of the past several hundred years, including such periods as the Little Ice Age. Moreover, the actual state of the deep ocean is difficult and expensive to determine.

Indeed the whole of this chapter illustrates why it is so difficult to predict the weather or to account for climate variation. Certainly the dominant factors are astronomical: the solar input and its distribution with latitude, the seasonal rhythms, the rotation of the Earth and its gravitational attraction. Still, the irregularities of the surface, continents, oceans and mountain chains, have profound influences. Important atmospheric phenomena like cyclones sometimes occur over very small distance scales (a few hundred kilometers or less) and can vary extremely rapidly. Thus a realistic model for computation[3] must involve a closely-spaced three-dimensional grid of points, to take account of all these spatial irregularities (mountains, coastlines, etc.) and to give some information on small-scale disturbances, and they must also be performed at short time intervals (days, hours, minutes). At the same time we noted that the state of the oceans depends on very long time intervals (centuries or millenia). Even with a relatively coarse grid spacing of 100 km horizontally and 500 meters vertically, we need at least 2 million points to represent the geosystem. The atmospheric or oceanic parameters must be evaluated in each of these points and their evolution followed. Something like 10^{10} operations mught be required to compute a day's evolution, using a time

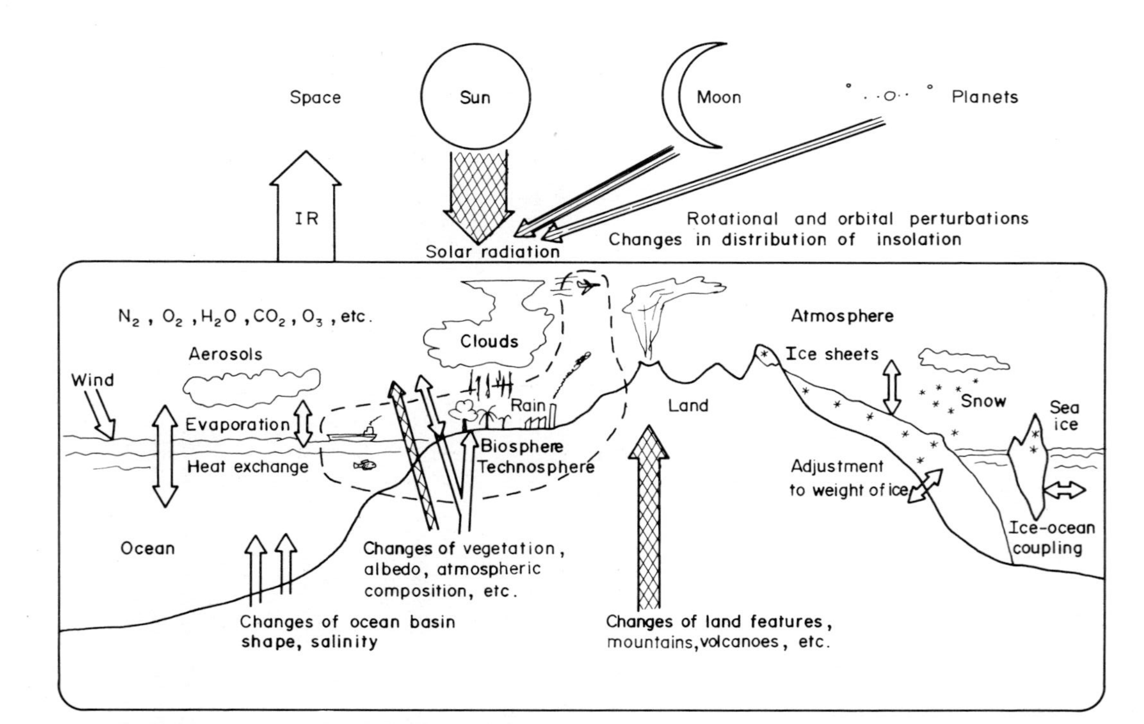

Figure 10-8. The climatic system and its environment. Hatched arrows indicate external processes, open arrows indicate internal processes in climatic change (after U.S. GARP Comm.).

step of 3 hours. On a very fast and huge computer requiring only a nanosecond per operation, this corresponds to 10 seconds of computation per day of weather. It would seem that to predict the weather a year in advance would require only an hour of computation. Understanding the results of the computation would probably take the research team much longer.

However, although this sort of numerical experiment is often used to study the influence of various factors on weather and climate, such a calculation could not possibly represent the weather except in a very gross way, since weather can vary over distances much shorter than 100 km, and once an error appears in such a calculation, it tends to perpetuate itself if not to grow. Serious numerical weather prediction, even for only a few days, requires a much finer grid. With a horizontal spacing reduced to 2 km and vertical spacing of 100 m, the number of points representing the geosystem would rise to $2.5\ 10^{10}$. For such a calculation to have any meaning, the time step used to follow evolution of the weather would have to be considerably reduced as well, say to 100 seconds. Then computing a day's evolution would require $2\ 10^{16}$ operations, taking several months at a nanosecond per operation. We could not keep up with nature! Obviously some simplifications are necessary, and even if computers become much faster still than they are today, it does not seem that "brute force" numerical computation will give us the answer. Human intelligence remains valuable in attacking the problem. Numerical experiments with more or less idealized models of the geosystem do give us some feeling for the ways in which the many factors determining climate interact. Real experiments with our global climate system are undesirable!

11

Life and the Earth

> The atmosphere, therefore, is the mysterious link that connects the animal with the vegetable, the vegetable with the animal kingdom.
>
> Dumas and Boussingault, 1844[1]

> From our point of view, therefore, the modern process of evolution of living organisms is fundamentally nothing more than the addition of some new links to an endless chain of transformations of matter, a chain the beginning of which extends to the very dawn of existence of our planet.
>
> A. I. Oparin, 1936
> *The Origin of Life**

WE have up to this point examined the various ways in which the cosmic environment and the terrestrial environment are related. Comparing the Earth with other planets of the solar system, we have seen how the laws of nature operate under different circumstances. The Earth is indeed radically different from the giant planets, and as we mentioned earlier in the book, these differences are fairly easy to understand. However, while there are many points in common between the Earth and the other "terrestrial" planets, especially Venus and Mars, the Earth has certain unique and remarkable properties. It is the only planet on which liquid water exists in quantity. Where Venus and Mars have essentially pure carbon dioxide atmospheres, one thick, the other thin, the Earth has an atmosphere of nitrogen and oxygen, not found on any other planet. Finally, the Earth appears to be the only abode of life in the solar system. These properties are interrelated. It is not just that life, once having established a foothold on Earth, has evolved adapting itself to changing terrestrial conditions. Life has in fact shaped these conditions. Indeed Margulis and Lovelock have suggested that "the Earth's atmosphere is regulated by life on the surface so that the probability of growth of the entire biosphere is maximized".[2]

*(transl. S. Morgulis, Dover Publications, Inc. 1953)

The present atmosphere of the Earth is certainly not its original, primary atmosphere, which may have resembled the present atmospheres of the giant planets. Such a reducing atmosphere, probably including, in addition to hydrogen and helium, such hydrogen compounds as methane, ammonia and water vapor, must have been lost in an early phase of the Earth's history. We have evidence for the loss of much material from the proto-Earth in the fact that neon, a cosmically abundant element (since ${}_8O^{16} + {}_2He^4 \rightarrow {}_{10}Ne^{20}$, as described in Chapter 4), is extremely rare on the present Earth, even though it does not escape easily under present conditions. This loss of the primary atmosphere may have been simply an effect of high temperature, but extreme activity in the young Sun with a strong solar wind may also have been involved. An any rate, the present atmospheres of the terrestrial planets must be at least secondary in nature, if not tertiary atmospheres following evolution of a secondary atmosphere. Gas is still escaping from the interior of the Earth today, as in volcanoes, so we can guess how a secondary atmosphere could have been formed; in the Earth's first eon of existence, when the crust was being formed, internal activity was probably much higher than today. Certainly the secondary atmosphere would not contain free oxygen, a highly reactive gas. Judging from Mars and Venus, as well as from present volcanic emissions, carbon dioxide must have been a major component; it is usually assumed that relatively inert nitrogen was important as well, and remembering the oceans we must assume that water vapor also emerged in large quantities from inside the Earth as part of the secondary atmosphere.

If this is our hypothesis, we must explain how each of the three planets evolved to its present state. The positive feedback of the runaway greenhouse effect proposed by Rasool and De Bergh seems to account for the present situation on Venus.[3] If Venus started off a bit hotter than the Earth because of its proximity to the Sun, much more of the water would remain in gaseous rather than liquid form. The greenhouse effect of the additional water vapor in the atmosphere would raise the temperature, accelerating evaporation, until all the oceans on Venus, if ever there were any, evaporated. The loss of the water vapor from the atmosphere must have occurred gradually through the escape of hydrogen, following splitting of the H_2O molecule by sunlight, but it must be admitted that it is hard to understand how this could account for the present extreme rarity of water vapor in the Venus atmosphere. Nitrogen, if any, must have been lost to the surface of the planet, and the oxygen left behind from the H_2O should be there too in the form of various oxides. The amount of CO_2 in the atmosphere is enormous, approximately 1000 tons per square meter. On Mars, by contrast, some of the CO_2 and H_2O may have escaped, and what is left is partially in frozen form or still locked up inside the planet; liquid water cannot now exist there. On Mars too the surface is oxidized. Since the atmosphere is very thin and contains virtually no oxygen or ozone, the surface is exposed to solar ultraviolet radiation, and this affects chemical processes there in various ways. As a consequence samples examined by the Viking lander simulated certain biological processes, although no biological material was found.

The Earth seems to have been located at just the right distance for most of the water vapor to have condensed and remain liquid, so that a runaway greenhouse was not triggered. Presumably, the conditions of the primitive Earth, involving liquid water, residual methane and ammonia from the rapidly disappearing primitive atmosphere, and energy sources including solar ultraviolet photons and atmospheric electrical discharges, were appropriate for the origin of life. It is known from experiment that complex organic molecules can be formed under these conditions. These extremely interesting questions are discussed in many books, and we shall not go into detail here. There is some evidence for the existence of microbial life at a very early epoch, about 3.3 eons B.P., well before there was much oxygen in the atmosphere.

The first life forms were surely anaerobic, obtaining energy from the process of fermentation, in the absence of oxygen. If we represent organic matter by the formula for glucose, this process can be given symbolically as

$$\underset{\text{glucose}}{C_6H_{12}O_6} \rightarrow 2\,\underset{\text{alcohol}}{C_2H_5OH} + 2\,CO_2 + \text{energy}$$

This is of course an idealization. Other elements, notably nitrogen, phosphorus and sulphur, are essential to life, and the energy indicated above is stored in the cell in the form of high-energy phosphates, such as adenosine triphosphate (usually abbreviated ATP, which is not a chemical formula). Thus carbon dioxide is a by-product of this anaerobic life. Such anaerobic life forms continue to exist on Earth, in oxygen-poor environments, playing an important ecological role.

At some point in the Precambrian, perhaps as early as 3 eons B.P., biological evolution "discovered" photosynthesis, a process by which solar energy is used to build up organic molecules. In algae and all higher plants, this involves rearranging the atoms of the CO_2 and H_2O molecules, in particular splitting the water molecule and releasing oxygen in molecular form as a by-product. The symbolic equation is

$$6\,CO_2 + 12\,H_2O + \text{photons} \rightarrow C_6H_{12}O_6 + 6\,H_2O + 6\,O_2$$

where the process is quite complex and requires the substance known as chlorophyll. With this process, given carbon dioxide in the atmosphere, water in the oceans, and sunlight, carbon can be "fixed" in organic matter at a high rate, much faster than is possible by the processes which presumably built up the first organic molecules. The mass of the biosphere can increase substantially. The problem, for the anaerobic life of these early times, is that photosynthesis pollutes the atmosphere, with oxygen, a poisonous gas. Certainly 2 eons ago, blue-green algae were excreting oxygen at a high rate. The appearance of free oxygen in the Earth's atmosphere is attested to by the "red beds" of fully oxidized iron corresponding to the late Precambrian, about 1 eon B.P.

The response of life to this pollution, caused by life itself, is further evolution. With the process of respiration, life devised a negative feedback process keeping the oxygen content of the atmosphere in check. Moreover, this process recovers solar energy, in the combustion of the carbohydrate in oxygen.

$$C_6H_{12}O_6 + 6\,O_2 + 6\,H_2O \rightarrow 6\,CO_2 + 12\,H_2O + \text{energy}$$

This process returns carbon dioxide to the atmosphere, making possible the approach to a steady state. Since photosynthesis leads to a great increase in the total biomass, and since respirative metabolism means that life can cope with oxygen pollution and indeed use it to release energy, the development of organisms of a new type becomes possible. These organisms, the heterotrophs, obtain their organic matter ready-made by feeding on other organisms, ultimately depending on plants, the autotrophs, which build up their organic molecules using photosynthesis and the solar energy input. This solar energy is then partially recovered by respiration, i.e. by burning the glucose in oxygen, as indicated in the last formula above. All in all, we see that the appearance of photosynthesis not only revolutionized the biosphere, it also led to the radical transformation of the Earth's atmosphere, to the present composition of 21% oxygen and 79% nitrogen. This transition must have been gradually accelerating, and substantially completed a few hundred million years ago. With oxygen in the atmosphere, ozone was formed, thus shielding all the Earth's surface from solar ultraviolet radiation. Life could emerge from under water and conquer the land.

It may seem incredible that a mere "scum" on the surface of the planet, with an average mass of only 6 kg/m^2, should so radically affect the Earth's atmosphere, whose mass is now over a thousand times greater, about 10^4 kg/m^2. We can however confirm this in various ways. The cycling of carbon between the land, oceans and atmosphere, is profoundly influenced by life. Nearly all of the carbon in the atmosphere is in the form of CO_2, of which there is about 2.4 10^{12} tons or 4.7 kg/m^2. Biological sources release — through respiration and decay — about 3.8 10^{11} tons of CO_2 per year, and they use up about the same amount in photosynthesis. Whether global balance holds or not depends on whether the global biomass is constant or changing, and on other processes, and we really do not know the present situation precisely. Human activities have an impact, which we shall discuss in the next chapter. At any rate the atmospheric CO_2 is cycled through the biosphere in a very short time, perhaps 10 to 15 years. Locally, when the air is calm, strong variations of CO_2 content are easily observed over a forest, respiration raising the CO_2 content near the ground at night, and photosynthesis reducing it, especially at treetop level, during the day. On a larger scale, strong seasonal variations are observed in the northern hemisphere (above 30° latitude), the CO_2 content dropping substantially over the growing season between April and November. It has been speculated that increases in the atmospheric CO_2 content due to Man's activities are in part limited by the negative feedback of increased plant growth. We should note however that the oceans contain substantially more CO_2 (in the form of bicarbonate ions) than either the atmosphere or the biosphere. Exchanges of CO_2 between atmosphere and ocean, depending on temperatures, surface waves, circulation of the oceans and atmosphere, etc., may be even more important. Finally, a still much larger amount of carbon is found in sedimentary rocks; exchanges between this enormous reservoir and both the oceans and atmosphere proceed very slowly, perhaps over millions of years, but of course they are extremely significant over the life of the planet. When we take into account the amounts of CO_2 contained in the sedimentary and oceanic reservoirs, we find a total of about 200 tons/m^2, still less than Venus, but comparable.

We have attributed the present oxygen content of the atmosphere to the action of photosynthesis. The atmospheric oxygen mass is presently about 10^{15} tons. Photosynthesis presently releases approximately 10^{11} tons per year to the atmosphere, and presumably the same amount is used in respiration and combustion of organic material. This implies that oxygen cycles through the biosphere in about 10 000 years. Again, much larger amounts of oxygen are found in the oceans, as sulphate ions, and

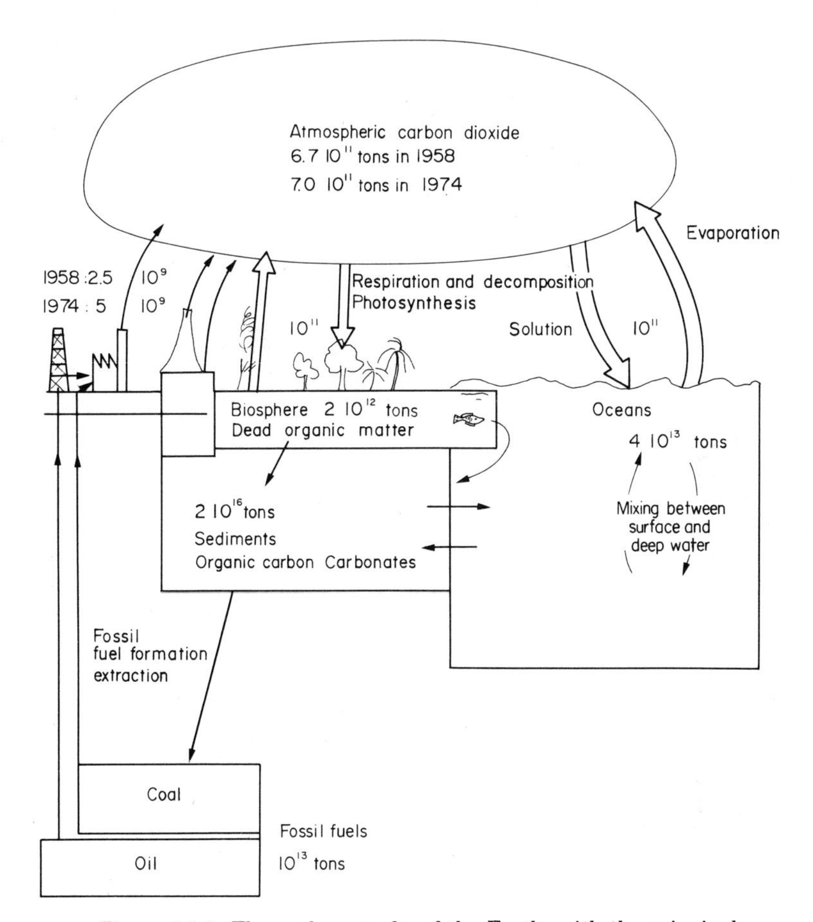

Figure 11-1. The carbon cycle of the Earth, with the principal reservoirs and fluxes indicated, with equivalent masses of carbon in metric tons, and tons per year. There is considerable uncertainty regarding all of these figures except for the atmospheric CO_2 and the rate of fossil fuel burning.[4]

especially as carbonate rocks and other "fossil" forms in sediments. The relative proportions of this total of oxygen and the total of carbon mentioned earlier, suggest that all of this was the result of photosynthesis operating over the last few eons, on an atmosphere of carbon dioxide initially comparable to that of Venus, in the presence of liquid water.[5]

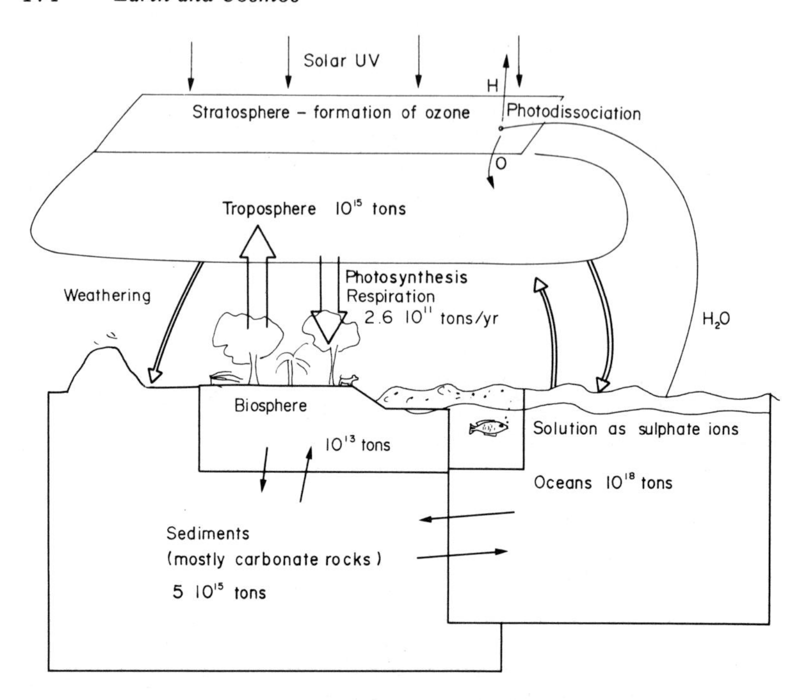

Figure 11-2. The principal elements of the oxygen cycle of the Earth. Note the overriding importance of the biosphere for the maintenance of oxygen in the atmosphere.

Nonbiological processes are much less effective. If life on Earth were suddenly to come to an end, the present atmospheric composition would not be maintained. The oxygen would gradually disappear from the atmosphere.

The major constituent of the Earth's atmosphere is nitrogen, a relatively inert gas. Not so long ago it was commonly assumed that Mars and Venus also had a substantial proportion of nitrogen in their atmospheres, but this does not seem to be the case. Nitrogen too is largely controlled by biological processes, just as it plays a controlling role in the biosphere. The atmospheric reservoir of nitrogen, about $4\ 10^{15}$ tons, is about ten times what is found in sediments, and very much larger than either the biospheric or hydrospheric reservoirs. However, most living things cannot use the molecule N_2 of the air directly. The nitrogen must be "fixed", i.e. it must be converted into a different form as part of various molecules, nitrites, nitrates, for example potassium nitrate, KNO_3. The existence of all living organisms on Earth depends for the most part on the ability of a relatively small number of micro-organisms to perform this operation, in various steps, and to release the fixed nitrogen in the soil or the ocean. The contribution of natural nonbiological processes is much smaller.

Once the nitrogen-fixing bacteria have done their work in the soil, the various forms of fixed nitrogen mount the food chain, returning to the soil when organisms die and decay. Since there is still a lot of nitrogen in the atmosphere, and since the nitric acid content of the oceans is still small, there must be processes returning nitrogen in molecular form to the air. These processes are dominated by the denitrifying bacteria, which return about 10^9 tons of nitrogen to the atmosphere each year. The cycling time of the nitrogen is thus very much longer than the cycling times of either oxygen or carbon dioxide, of the order of millions of years. In the absence of life, the nitrogen content would decrease very slowly, as a result of natural nonbiological nitrification processes, such as oxidation of nitrogen in lightning flashes. Over a planetary lifetime of a few eons, all the nitrogen might disappear, as seems to have occurred on Venus. On the planet Earth, in the presence of Man, the nitrogen cycle has been strongly perturbed, as we shall see in the next chapter.

We have discussed the role of life in maintaining the abundance of the principal atmospheric constituents, N_2, O_2 and CO_2. The abundances of many trace constituents are also biologically controlled. It is striking that many gases and aerosols which we would call pollutants are absorbed or excreted by many different life forms. Many more metabolic processes exist in nature than the few major ones we have cited here. Thus the present methane (CH_4) in the Earth's atmosphere is the product of the putrefaction of organic material as a result of the action of anaerobic methane bacteria. About $2\ 10^9$ tons are released each year to the atmosphere, which contains only about $3.6\ 10^9$ tons, so that the cycling time is about 2 years. Sometimes this burns at the surface, as in the *ignis fatuus*, but most of the methane is probably oxidized to CO_2 and H_2O in the stratosphere. This high altitude water vapor is then broken up into hydrogen and oxygen by the solar ultraviolet radiation, and the hydrogen escapes to space while the oxygen remains behind. In this way these bacteria may contribute indirectly to the supply of oxygen in the atmosphere. The total oxygen content can probably not get very much higher than its present value, without provoking a substantial increase in the frequency of natural forest fires, a stabilizing, negative feedback mechanism. This is important because oxygen in excessive concentrations remains a menace to living things, even to oxygen-breathing organisms.

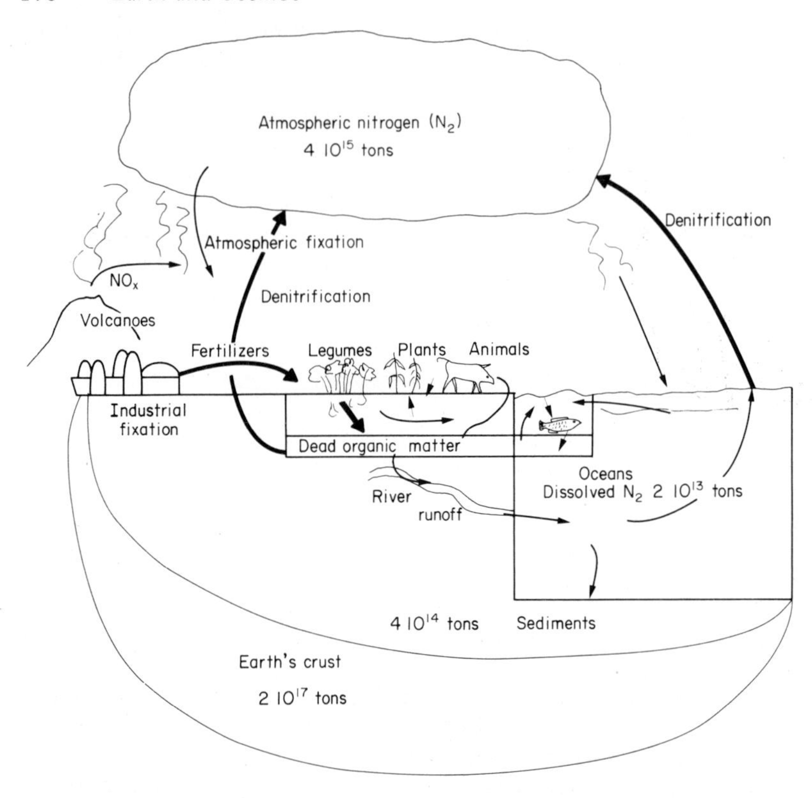

Figure 11-3. The principal elements of the nitrogen cycle on the Earth. Human intervention in the fixation of nitrogen is important. All figures except for the atmospheric content are extremely uncertain.[6]

We have seen how the processes of life maintain the present composition of the Earth's atmosphere, and how indeed they must have transformed it. The radical transition from a reducing atmosphere to an oxidizing one was certainly accompanied by enormous ecological transformations, both inducing and being induced by rapid biological evolution. The role of the smaller organisms, particularly microorganisms, has been crucial, and these life forms are precisely those in which evolution can proceed rapidly by natural selection in response to changing environmental conditions.

Lovelock and Margulis, in what they call the "Gaia hypothesis",[2] suggest that the "complex entity involving the Earth's atmosphere, biosphere, oceans and soil . . . constitutes a feedback or cybernetic system which seeks an optimal physical and chemical environment for the biota". They suggest that various biological processes, including those we have just described, control not only the atmospheric composition (and so the greenhouse effect), but also the surface albedo and emissivity, the amount of atmospheric aerosols, acidity of water, electron concentration, and the circulation of elements such as sulphur and phosphorus, keeping conditions on Earth within limits tolerable to life. They argue that external factors, notably the Sun's luminosity, have varied far more over the Earth's history than has the surface environment, even considering the Ice Ages, which they describe as partial but not total failure of the control system. They note that climate models, which do not include biological processes, generally predict an extremely unstable situation, with only a thin margin between a runaway greenhouse and a completely ice-covered planet. If in the remote past the greenhouse effect kept the Earth warm and unfrozen when the Sun was less luminous, why did it not run away as the Sun brightened? The drop in CO_2 due to the proliferation of photosynthetic organisms seems to have come at just the right time. It is not suggested that some planetary engineer was consciously adjusting the system, but rather that the mechanisms of natural selection operating locally, interact to form a planetary environmental control system.

The "Gaia hypothesis" is an extremely provocative way of looking at the problem of life on the Earth, to which all sorts of investigations in different fields can be referred. The crucial questions may be whether or not it is true that, in the absence of life, the external factors would produce climate catastrophes that could not be avoided by nonbiological feedback mechanisms. The present climate models are still crude, and the existence of oceans may be sufficient to ensure stability. One may in any case wonder how the suggested biological feedback mechanisms can be so successful, i.e. how they can "know" just what to do. The answer may be that if they had failed, we wouldn't be here! This leaves open the possibility of a future failure, and as the principal actors on the current biological scene, we humans should reflect on what we are doing.

12

The Impact of Man

> Nature has no human inhabitant who appreciates her. The birds with their plumage and their notes are in harmony with the flowers, but what youth or maiden conspires with the wild luxuriant beauty of Nature? She flourishes most alone, far from the towns where they reside. Talk of heaven! ye disgrace earth.
>
> Henry David Thoreau[1]
> *Walden* (1854)

FOR a distant astronomer, the very existence of the planet Earth would have been hard to establish until very recently, when technological society signaled its existence, and that of the Earth, by the emission of radio waves. Over the past few decades, the power emitted by the Earth at wavelengths of a few meters has increased by an enormous factor, reaching a level comparable to that of the Sun at these same wavelengths. Even without decoding the television broadcasts involved, our remote radioastronomer might be able to deduce from the variations in the overall intensity of these emissions and from the shifts in specific spectral lines (stations operating at specific frequencies), that these radio waves come from a nonuniform distribution of emitters on the surface of a rotating object moving around the star we call the Sun. By decoding the transmissions, he or she might learn much more about our planet and our society, although it would certainly be confusing.

This is the extent of our present impact as seen from very far away. From closer up, with increasing resolution of details on the surface, and more information on the composition of the atmosphere, other activities of the human species become apparent. That Man, constituting less than one-tenth of one per cent of the mass of the biosphere, itself a mere scum on the surface of the planet, can have a global impact, may seem remarkable. However, the first photosynthetic organisms succeeded in transforming the whole terrestrial environment, much more radically than Man has done to date, and so the idea of a global impact by *homo sapiens* should not be surprising.

There has been much discussion in recent years of the pollution of the environment by Man, with attention mostly being given to industrial activities. However, the impact of Man on the global environment goes back to the agricultural revolution, about 10 000 years ago. Before that, humans were simply one omnivorous species among many, hunting and gathering for sustenance, with hardly any organized efforts to change the environment. Even so, the idea of an ecological equilibrium is illusory; as a particularly effective predator, Man was no doubt responsible for the extinction of many species of large mammals, particularly in the western hemisphere. Environmental conditions varied, notably through the ice ages, and Man seems to have extended his range and increased his population throughout these periods. It seems likely that the world human population did not reach 10 million in pre-agricultural times. The rapid rise in the human population, and the corresponding increase in its environmental impact, are not the result of biological evolution, but rather that of social and cultural evolution. With the invention of agriculture, Man began an enterprise of deliberate transformation of the biosphere, involving large modifications in the numbers of certain species, i.e. cultivated plants and domesticated animals. This led to enormous growth of the human population, which reached 800 million in 1750. The average human population growth rate over the prior 10 000 years, less than 0.1%, may seem low by modern standards, but this growth rate could not have been sustained in the absence of agriculture. In North America, where except in Mexico and parts of the southwest United States, agriculture had not been adopted before the 15th century, the arrival of Europeans (and the forced immigration of Africans) led to an enormous increase in the population, accelerated of course by the industrial revolution, but nevertheless primarily due to the conversion of the new land to agriculture.

The global impacts of the agricultural revolution are multiple. The microclimate is certainly changed wherever land is converted to agricultural use: the local surface albedo is modified and the local water budget is changed, both through direct evaporation where irrigation is used, and through the differences between evapotranspiration by the wild and cultivated plant species.[2] At present about 10% of the total land surface of the globe is devoted to agriculture, and it should be noted that this constitutes nearly a third of the land that can support vegetation. It is quite likely that changes in land use and vegetation have affected the CO_2 budget which we discussed in the preceding chapter, but it is hard to say how much. George Woodwell and his coworkers at the Woods Hole Oceanographic Institution argue that deforestation in tropical regions is proceeding so rapidly that the rate of CO_2 release to the atmosphere is several times higher than that caused by the burning of fossil fuels.[3] Many other researchers disagree; they believe that both deforestation and the *net* CO_2 release are much smaller, with new growth fixing carbon quite rapidly. If Woodwell is right, it is hard to understand why the atmospheric CO_2 content is not rising even more rapidly than observed. Many, but not all, agronomists believe that a rise in atmospheric CO_2 enhances plant growth and photosynthesis on a global scale, thus providing a negative feedback in the carbon cycle. However, we are far from being able to evaluate changes in global biomass with sufficient accuracy to check this.

Agricultural practices have also very seriously affected the nitrogen cycle, even before the introduction of artificial fertilizers. Certain crops known as legumes, such as soy beans or alfalfa, exist in symbiosis with specific varieties of nitrogen-fixing bacteria, much more effective than free-living varieties. The cultivation of such crops, partly for their own nutritional value for humans or cattle, partly to return fixed nitrogen to the land, substantially increases the rate of fixation of nitrogen in the soil, locally by factors of 100, globally perhaps by as much as 50%. With the introduction of industrial fertilizers in large quantities over the last few decades, the rate of nitrogen fixation has nearly doubled globally. This is a huge perturbation, but most of it is absorbed by the negative feedback of the biosphere, as denitrifying bacteria have multiplied in response to the opportunity given them. Still, it appears that a net excess of about 9 million tons of nitrogen are fixed per year. This is no threat to the huge nitrogen content of the atmosphere, but it does imply a build-up of fixed nitrogen in the soil, rivers, lakes and oceans. Moreover, some of the fixed nitrogen is returned to the atmosphere by the denitrifying bacteria in the form of N_2O, and if it reaches the stratosphere, it may affect the ozone content, as we shall see later.

Agricultural land use may have still another effect on the global environment. Many of the particles (aerosols) in the troposphere are known to come from deserts. Thus dust from the Sahara often crosses the Atlantic and is observed in Barbados. In northwest India, the Rajasthan desert is a source of dust storms; the treeless loess-covered plateaus of northern China are a source of dust over China and Japan, and part of the North Pacific. Now some of these deserts, in particular the Rajasthan desert of India, certain areas of the Middle East (ancient Mesopotamia), and parts of North Africa, appear to be the result of poor agricultural practices. Certainly this was the case for the American dust bowl of the 1930's. Desertification, the spreading of deserts over previously productive land, is a matter of great concern to many nations of the Third World, for obvious reasons, since it affects the carrying capacity of their own territory. As a factor increasing the global aerosol load of the atmosphere, it affects us all.

Another important contribution to atmospheric aerosols is the smoke from agricultural burning. Such slash-and-burn agriculture not only affects the CO_2 budget, it also produces a blue haze over areas in Brazil, central Africa, and southeast Asia. It is estimated that over 160 000 km^2 of natural forest are thus destroyed each year in the Third World countries. Since evapotranspiration in tropical forests is an important factor in the global water budget, a continuance of this trend could have effects on the global circulation pattern as well as on aerosols in the atmosphere and on the CO_2 budget.

Atmospheric dust is, as we have seen, partially of agricultural origin. With the industrial revolution, a new source of air pollution appeared on a large scale. Most coal, and much petroleum, contains significant amounts of sulphur, and when these fossil fuels are burned, the products include not only soot particles, but also gases such as sulphur dioxide (SO_2) and various hydrocarbons which become aerosols after a short time in the atmosphere. These particles are emitted over the land, mostly over the industrially developed countries of the northern hemisphere. How far they move depends on their sizes, the atmospheric circulation, and scavenging processes such as rain and snow which return them to the surface. Obviously these particles do not respect national boundaries, and the sulphates emitted in England and Germany often fall as sulphuric acid rain in Sweden. Still, most of the larger particles do not remain in the atmosphere for more than a week, so that the polluted areas are not very far from the emitters of the pollution. It does not seem unfair to relate "Gross National Pollution" to Gross National Product, where industrial pollution is concerned.

Just how important are these anthropogenic particulate emissions, compared to natural sources of aerosols? Locally, there is no doubt that their impact is often large. Such measures of atmospheric particle load as the rate of dust fall in the Caucasus, or the number of smoke-haze days in Chicago, correlate relatively well with indicators of industrial activity. On the other hand, in other locations such as Washington DC, regular seasonal fluctuations point to a natural origin of many of the aerosols. The highest large-scale aerosol concentrations in the United States are found over the Blue Ridge Mountains in Virginia, where the blue haze is the result of the interaction of sunlight with certain hydrocarbons called terpenes emitted by trees.

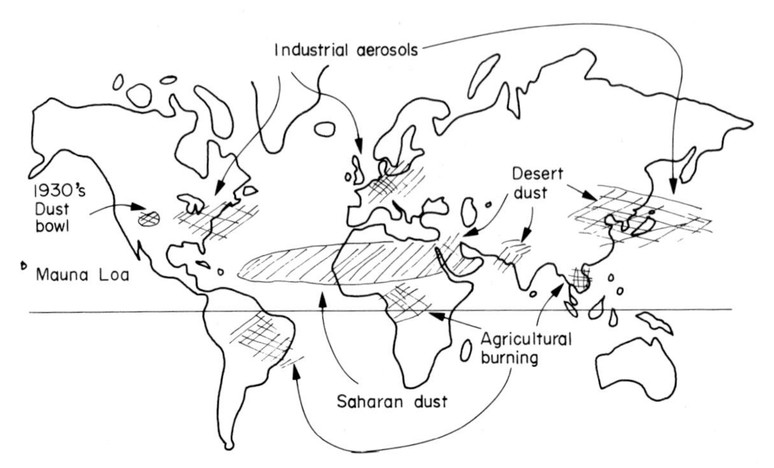

Figure 12-1. Natural, agricultural and industrial aerosols. Distribution in the troposphere.

One way of estimating the total atmospheric aerosol load is to measure the atmospheric *turbidity*, i.e. the fraction of energy removed from the direct solar beam under cloudless conditions. Such measurements have been made monthly for the last 20 years at Mauna Loa, on Hawaii, far from all sources of industrial and agricultural pollution, thus presumably representing some sort of global average of the atmospheric aerosol content. These measurements show fluctuations which can best be explained in terms of strong injections of aerosols during volcanic eruptions, notably that of Mt. Agung in 1963, followed by gradual fall-out of the particles. There is very little sign of anthropogenic influence. However, the Mauna Loa Observatory is located at an altitude of 3500 meters. Violent volcanic eruptions easily send material into the stratosphere, but apart from atmospheric nuclear explosions and high-flying aircraft, nearly all anthropogenic emissions occur at the Earth's surface, into the troposphere. Satellite observations show very long pollution plumes extending from major point sources of pollution, like the Black Mesa power plants in the Four Corners regions of the southwest United States. Although generally there is mixing vertically to altitudes well above 3500 m in the troposphere, this can be hampered by the formation of a *temperature inversion layer*, i.e. a layer in which the temperature is higher than at the surface. This is rather common over the Pacific, because of the high rate of evaporation at the surface. Thus the Mauna Loa observations do not prove that anthropogenic aerosols are negligible on the global scale. Similarly, the

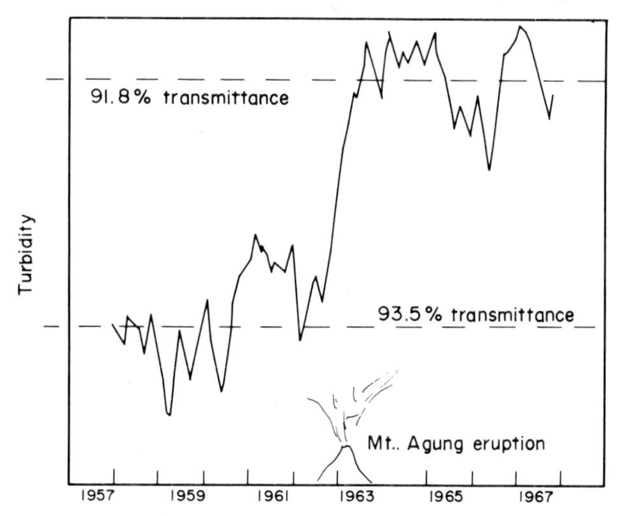

Figure 12-2. Atmospheric turbidity as measured at the Mauna Loa Observatory, illustrating the impact of the Mt. Agung eruption.

absence of noticeable effects in Antarctica may simply be a consequence of its remoteness from nearly all anthropogenic sources of particles.

All in all, it appears that the global *stratospheric* aerosol load is presently dominated by the sporadic input from volcanic eruptions. At *tropospheric* levels, there are many important natural sources, notably sea-salt spray, volcanic sulphates, and windblown dust from natural deserts. The estimates vary widely, but it appears that industrial release of SO_2 is now about the same as the average volcanic release, about 200 million tons per year, while agricultural burning releases about 60 million tons per year. The total particle release, both natural and anthropogenic, is somewhere between 1000 and 4000 million tons per year. Thus the contribution of Man is not negligible, but neither is it dominant, on a global scale.

In any event, the effects of aerosols on climate are multiple, and the result is not clear. Aerosols can scatter solar radiation, contributing to the global albedo and thus to global cooling, and this has often been assumed to be the major effect.[4] However, depending on their nature, these particles can also *absorb* solar radiation, increasing heating at the atmospheric levels at which they are found, and both scatter and absorb infrared radiation, contributing to the infrared opacity of the atmosphere and thus to the greenhouse effect. Under certain conditions, the net effect is one of warming. To evaluate correctly the impact of anthropogenic particle emissions on climate, we have to examine their properties, their size distribution, and their distribution with height in the atmosphere as well as over land and sea. While it seems likely that stratospheric aerosols mostly of volcanic origin, lead to global cooling at the Earth's surface, and may be responsible for the global cooling trend of the last 30 years, the anthropogenic tropospheric aerosols probably have a net *warming* effect.

The impact of Man on the present global particulate loading of the atmosphere is thus still ambiguous. By contrast, it is very well established that we are at present changing the composition of the atmosphere, increasing its CO_2 content by the combustion of fossil fuels. This appears clearly in the measurements of the atmospheric CO_2 content made at the Mauna Loa Observatory. The annual average has risen from 315 parts per million by volume in 1958 to 331 ppm in 1974, and this trend has been confirmed by measurements made in other locations. About a century ago, the CO_2 content appears to have been 294 ppm. This clear trend, a rise of 10% in the past century, accelerating in recent years, is almost entirely the result of human activities. The influence of deforestation and agricultural practices is uncertain. However, the amount of CO_2 released by the burning of fossil fuels can be estimated. The total, from 1860 to 1979, is believed to be 0.63 10^{12} tons, whereas the actual atmospheric content will have increased from 2.23 to 2.57 10^{12} tons.[6] Thus about half of the industrial CO_2 production has remained in the atmosphere, and half must have been absorbed in the oceans. Thus mankind has radically modified the global carbon cycle. We are rapidly depleting the fossil fuel reservoir at a rate many thousands of times faster than the natural processes which build up fossil fuels from the biomass. In the current decade alone, fossil fuel combustion will have produced 0.18 10^{12} tons of CO_2, and roughly half of this will have been added to the atmosphere. Here the technological evolution of Man clearly affects the Earth's atmosphere on a global scale.

As already mentioned, the importance of carbon dioxide for climate resides in its contribution to the greenhouse effect, since it absorbs in the infrared, mainly between 12 and 18 μm wavelength. It is not the major contributor, that being water vapor, which absorbs at infrared wavelengths from 5 to 8 μm and beyond 19 μm. The direct effect of an increase in atmospheric CO_2 can be evaluated, and it is not negligible. Various global model calculations suggest that a doubling of the CO_2 concentration should produce a global average temperature increase of 2–3°K. Of course we have at present "only" increased the CO_2 content by 10%, and this would correspond to a global warming of a small fraction of a degree

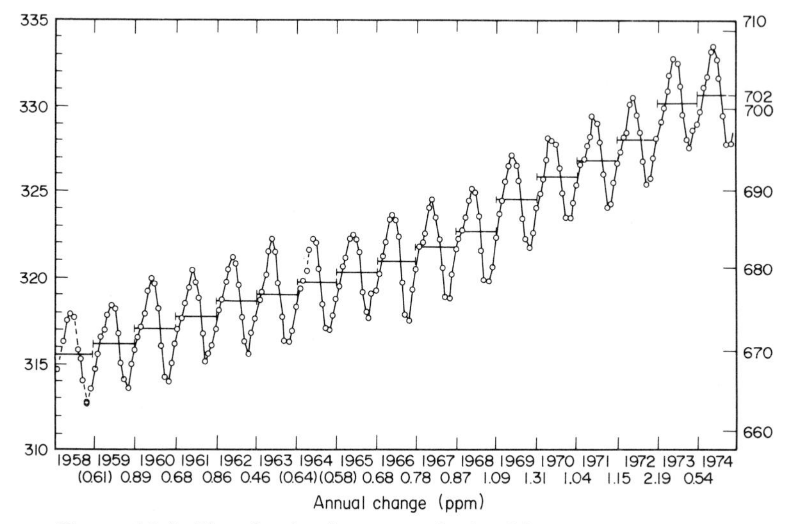

Figure 12-3. The rise in the atmospheric CO_2 content over the last two decades. The scale to the left gives the CO_2 concentration in parts per million by volume. The scale to the right gives the carbon mass in the atmosphere, in units of 10 million tons. Note the regular seasonal oscillations superposed on the long-term trend. Data from the Mauna Loa Observatory.[5]

since 1860, which may have very little to do with the actual warming trend which apparently ended in the 1940's. An important question is whether the global warming associated with CO_2 increase is reinforced by positive feedback or limited by negative feedback from the climate system. The fact that increasing the temperature of the oceans should lead to increased release of CO_2 and water vapor to the atmosphere is an element of positive feedback; the extreme is the runaway greenhouse effect discussed for Venus, but this is not possible on Earth before a few billions of years.

The warming associated with a rise in CO_2 in the air may trigger still other positive feedback mechanisms. We have already mentioned that the Arctic Ocean ice is relatively vulnerable, and if greenhouse warming led to its disappearance, further changes in Arctic climate would be expected, affecting the global circulation and climatic belts. There may however be negative feedback in the system. Some can be supplied by enhanced biological growth limiting the CO_2 increase. The problem is that other human activities and different types of pollution (DDT, oil films affecting phytoplankton, for example) may damage these capacities of the biosphere.

Enhanced evaporation and cloudiness associated with warmer temperatures may also be a major negative feedback mechanism, but we still have much to learn about cloud physics before we can have much confidence in the model results. There may be other unsuspected feedback loops. We cannot be absolutely certain of the global effect of increasing carbon dioxide in the atmosphere, although it is very probably a warming effect; calculations of the specific effects in various regions are not yet reliable. However, it would surely be foolhardy to assume that we are not already inadvertently modifying the climate.

We have discussed various forms of environmental pollution which have an indirect effect on the Earth's energy budget and thus on climate, through their effects on the global albedo and on the atmospheric absorption of radiation, on the infrared opacity of the atmosphere, and on processes involving water. There is however a possibility of direct perturbation of the energy budget through the release of heat, often called *thermal pollution.* Actually two phenomena are involved here. Most industrial processes, whether they be the production of steel, aluminium or electricity, have limited "First Law" thermodynamic efficiency; waste heat is produced and must somehow be dispersed in the local environment. In the United States, most electricity is generated by thermal power plants using fossil or nuclear fuels. Only 31.5% of the energy content of these precious nonrenewable resources is actually converted into the desired electric power, the 68.5% remaining being waste heat, released to the local environment and often having undesirable ecological effects as when the temperature of river water used for cooling is raised. In Sweden, again depending mostly on fossil and nuclear fuels, the fraction converted into electric power is actually lower (29%) but part of the heat output (24% of the energy input) is used for district and industrial heating. Still nearly half (47%) of the energy input ends up as waste heat.[7]

Where industrial processes such as iron and steel production are concerned, the energy actually consumed is again several times the amount strictly necessary to perform the chemical changes involved. This assertion is based on the First Law of Thermodynamics, which says that while energy can change form, it must be conserved. Still, to call the process wasteful is somewhat unfair. Although by the First Law it is theoretically possible to convert *all* the input energy into the desired form, as for example when smelting iron, according to the Second Law of Thermodynamics, such efficiency is only possible when the conversion proceeds infinitely slowly. In the real world, our time is measured, and we pay for it by degrading some of our input energy (and usually most of it) into waste heat.

Indeed, life itself can be described as a "negentropy" process, producing and maintaining an ordered and at times growing structure at the expense of its environment. We are all thermal polluters! However, until recently, our thermal pollution was limited to the value defined by our metabolism, about 100 watts per person, or on the global average only 1 milliwatt/m^2 even for today's population of 4 billion. This is based on the food we eat and this in turn depends on recent absorption of solar energy in the biosphere, except to the extent that fossil fuel inputs (industrial fertilizers, fuel for tractors, etc.) are needed to produce this food.

Nearly all the energy content of the fossil and nuclear fuels we consume ends up as heat, even that fraction that has done what we wanted done. Except for rockets leaving the Earth, all the energy used in transportation ends up as waste heat, released either in frictional deceleration and braking (there is no point in going somewhere if you can't stop), or in the dissipation of noise and turbulent air motions provoked by the moving vehicle. Of course energy consumption of transportation systems could be reduced by using electromagnetic braking or flywheel energy storage systems to recover some of the kinetic energy which must be lost during deceleration, but such systems are relatively rare. Most artificial lighting, except for city lights that escape to space (see Fig. 12-8), also leads to heating as the light is absorbed. The power needed to run machines corresponds to the heat generated by friction and noise, for the most part. Only the energy needed to perform chemical changes according to the First Law, as in producing aluminium from bauxite, does not rapidly end up as heat in the geosystem. All in all, we see that wherever we use energy, we have thermal pollution.

Thus apart from major power plants located in sparsely populated regions, generating true waste heat, the principal sources of thermal pollution on Earth are the densely populated highly urbanized regions of industrial activity in the developed countries. On a global average, about 100 watts/m^2 of solar energy are absorbed at the Earth's surface. Also on a global average, anthropogenic thermal pollution is much smaller, about 20 milliwatts/m^2. However, on a local level the rate of energy release is sometimes comparable to or even higher than the solar average. In industrial cities of northern latitudes, the anthropogenic energy release often substantially exceeds direct solar input during the winter months, when of course solar input is lowest and the need for heating is highest. The relative magnitude of such thermal pollution decreases as larger areas are considered. On Manhattan Island, in New York City, 630 W/m^2 are consumed, and released, compared with 93 W/m^2 of natural solar energy. In Moscow, we have 127 W/m^2 as compared to 42 W/m^2 of solar flux.

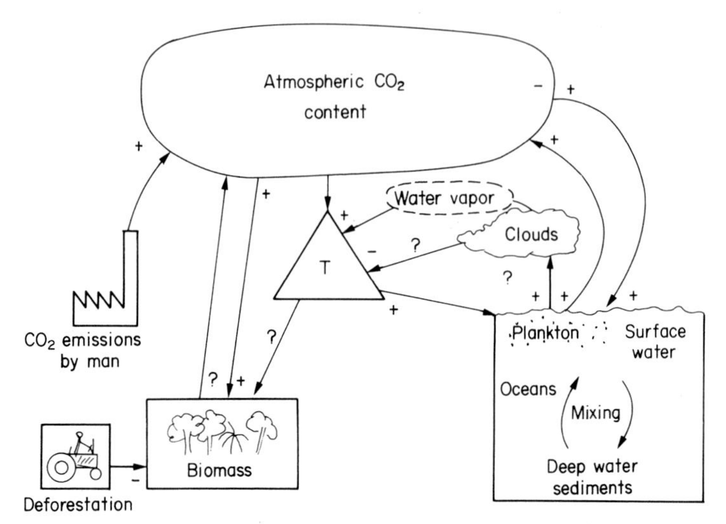

Figure 12-4. Part of the complex web of relationships between human CO_2 emissions, the biosphere, the oceans and climate. Thus the oceans may supply positive feedback to any greenhouse warming by CO_2, since with such warming they release more CO_2 and water vapor to the atmosphere. On the other hand, there is negative feedback from an increase in cloud cover. The biosphere (essentially photosynthetic plant life) should supply negative feedback to any CO_2 increase, unless the associated warming reduces biomass. Deforestation constitutes an additional perturbation by Man.

On the other hand, when we consider the entire Washington-to-Boston urbanized strip of the northeast United States, the average energy release falls to 4.4 W/m^2. For the United States as a whole, the figure is 0.24 W/m^2, but for the German Federal Republic it reaches 1.4 W/m^2, and there the solar input only averages about 50 W/m^2. These estimates were all made in the late 1960's.[9]

Clearly, global thermal pollution is not at present significant, representing only about 0.01% of the average absorbed solar flux. However, local effects can be quite pronounced. The "heat island" effect for cities is well known. There are some indications of such effects reaching a regional scale in densely industrialized areas of the globe, notably in the United States. When such regional heat islands form together with temperature inversions inhibiting vertical circulation of the air, they lead to extremely unpleasant periods during which the cities enclosed must "stew in their

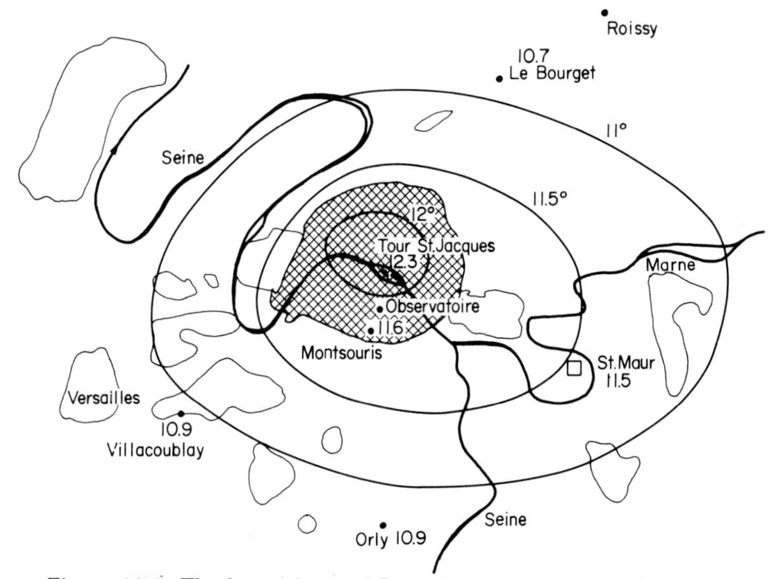

Figure 12-5. The heat island of Paris. Mean annual temperatures. The city lies in a basin surrounded by only rather low hills. (after Dettwiller[8]).

own juice" of pollutants and extra heat. Such stagnant air masses can divert regional air flows. It is not clear at what point such perturbations of the regional atmospheric circulation will begin to affect the global circulation pattern.

Much of our discussion of the present impact of Man has centered on climate. We have mentioned Man's agricultural engineering of the biosphere, but we shall not go into detail concerning the consequences of such pollutants as DDT except to note that many of our interventions of this type seem to provoke effective response by rapidly evolving insects and microbiota, rarely in ways favorable to our continued proliferation as a species. We shall conclude this chapter with a discussion of the ozone problem, where our modification of the atmosphere may lead to an undesirable change in the radiation environment. We recall that ozone is formed in atmospheric layers between 15 and 50 km altitude, and that this molecule absorbs the solar near ultraviolet radiation at wavelengths from 280 to 320 nm. Although the total energy flux in this wavelength range is small, such photons individually have energies which can produce physiological effects. In particular, such near-UV irradiation seems to be

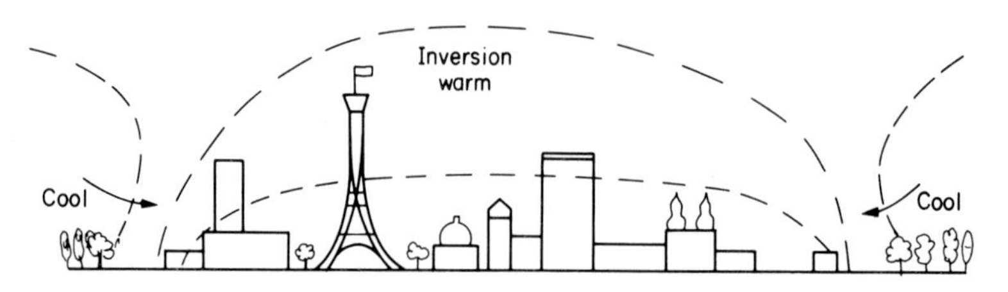

Figure 12-6. The heat dome above the city on a windless night.

linked to skin cancer, and this is one reason for the great interest in this problem. It was felt for a long time that the ozone layer of the atmosphere was safe from human tampering, and serious consideration of possible human impact only began when regular commercial air traffic in the ozone layer was envisaged. As noted earlier, the absorption of the solar UV in these layers leads to heating, so that the temperature increases with height, and such a stratification is relatively stable. To a first approximation, the stratospheric layers do not mix with the great mass of the troposphere, and therefore pollutants emitted directly in the stratosphere will not be rapidly diluted as much as they would be lower down in the troposphere.

Strong research efforts were launched in several countries, especially the United States, in order to evaluate the effects of pollution by large numbers of supersonic aircraft flying in the stratosphere. The major products of combustion of jet fuel are water vapor and carbon dioxide, but these would have little effect on the chemistry of the stratospheric layers. Depending on the sulphur content of the fuel, more or less sulphur dioxide may be emitted, leading to the formation of sulphate aerosols. This could increase the global albedo and lead to global cooling, but it seems unlikely that the input from supersonic aircraft could even remotely approach the natural input from volcanic eruptions. Moreover, the sulphur content of fuels for stratospheric jets could be regulated.

Of the other pollutants emitted by jet engines, the ones that can most affect the ozone content of the stratosphere are the oxides of nitrogen, NO and NO_2, often abbreviated as NO_x, the rate of emission depending on engine technology and not on the fuel used. These molecules can destroy ozone through the reactions:

$$NO + O_3 \rightarrow NO_2 + O_2$$
$$NO_2 + O \rightarrow NO + O_2$$

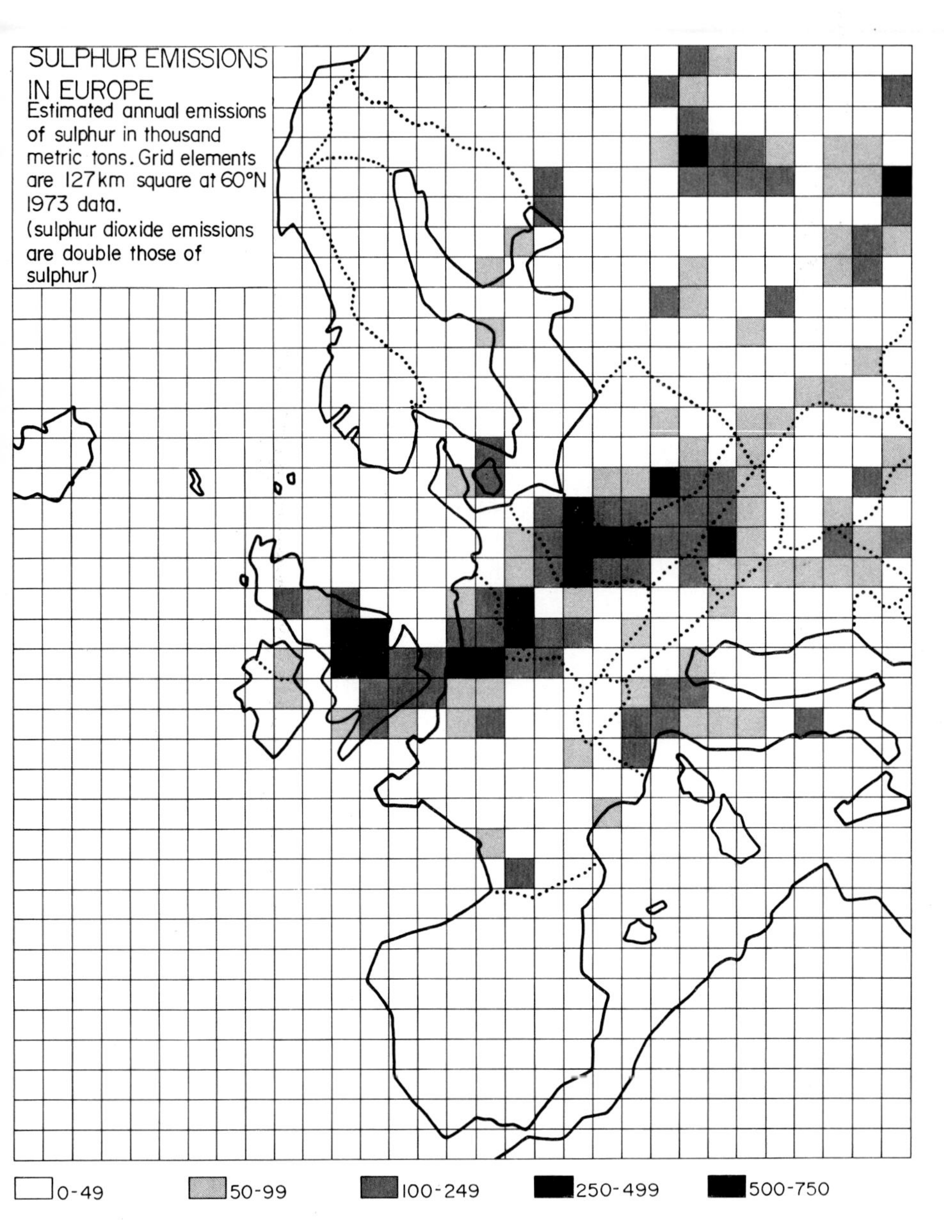

Figure 12-7. A regional problem. Industrial emissions of sulphur (mostly in the form of sulphur dioxide gas) lead to the formation of both sulphate aerosols and sulphuric acid in rain. Germany, the United Kingdom, France and the Benelux countries are major exporters of pollution. The Scandinavian countries and Finland are unwilling importers. Reprinted from the O.E.C.D. Observer/ No. 88 — September 1977.

which have the net effect of converting "odd oxygen", i.e. both atomic oxygen and ozone, into the O_2 molecule, without the NO and NO_2 molecules being destroyed in the process. Obviously this does not lead to the complete destruction of ozone; there are many more chemical reactions, some of them involving solar radiation, in the stratosphere, and a full calculation must take into account the interaction of all of these. Also, there are natural sources of the nitrogen oxides, such as volcanic eruptions, and ionospheric formation of NO by the action of solar X-rays, or solar protons; some of these natural inputs of nitrogen oxides can cause large natural variations of the ozone layer. Our knowledge of the concentrations of the various trace gases in the stratosphere and indeed of ozone itself was very incomplete when these studies were started only a few years ago, but much progress has been made.

Given the very small present world supersonic transport fleet, one can hardly assert a current impact on the stratosphere from such activities. However, the research stimulated by this problem did reveal that the stratospheric ozone content could be altered by emissions of trace gases. Some defenders of the SST noted that it was not realistic to assume that all emissions in the stratosphere remain confined and undiluted there. There is some mixing between stratosphere and troposphere, particularly as the tropopause altitude varies with the seasons and with latitude. This does indeed dilute pollution by stratospheric aircraft. However, it also raises the possibility that tropospheric pollutants can reach the ozone layer. Very large numbers of subsonic aircraft, such as the Boeing 707 and 747, or the McDonnell-Douglas DC-10, presently release more NO_x than even 10 000 Concordes. With mixing possible between stratosphere and troposphere, even ground-level pollution can affect ozone. In particular, it was noted that chlorine also can destroy odd oxygen, through the reaction chain:

$$Cl + O_3 \rightarrow ClO + O_2$$
$$ClO + O \rightarrow Cl + O_2$$

and that stratospheric concentrations even as small as 10^{-9} (1 ppb) could deplete ozone by 2%. Some stratospheric chlorine (in the form of CH_3Cl for example) is of natural origin, but anthropogenic contributions are significant. These involve carbon tetrachloride (CCl_4) and especially the chlorofluorocarbons (mainly $CFCl_3$ and CF_2Cl_2) used in refrigeration and as propellants in aerosol spray cans. When these gases reach the stratosphere, they yield free chlorine by absorption of solar ultraviolet (175—220 nm) photons, and this chlorine then destroys ozone as indicated

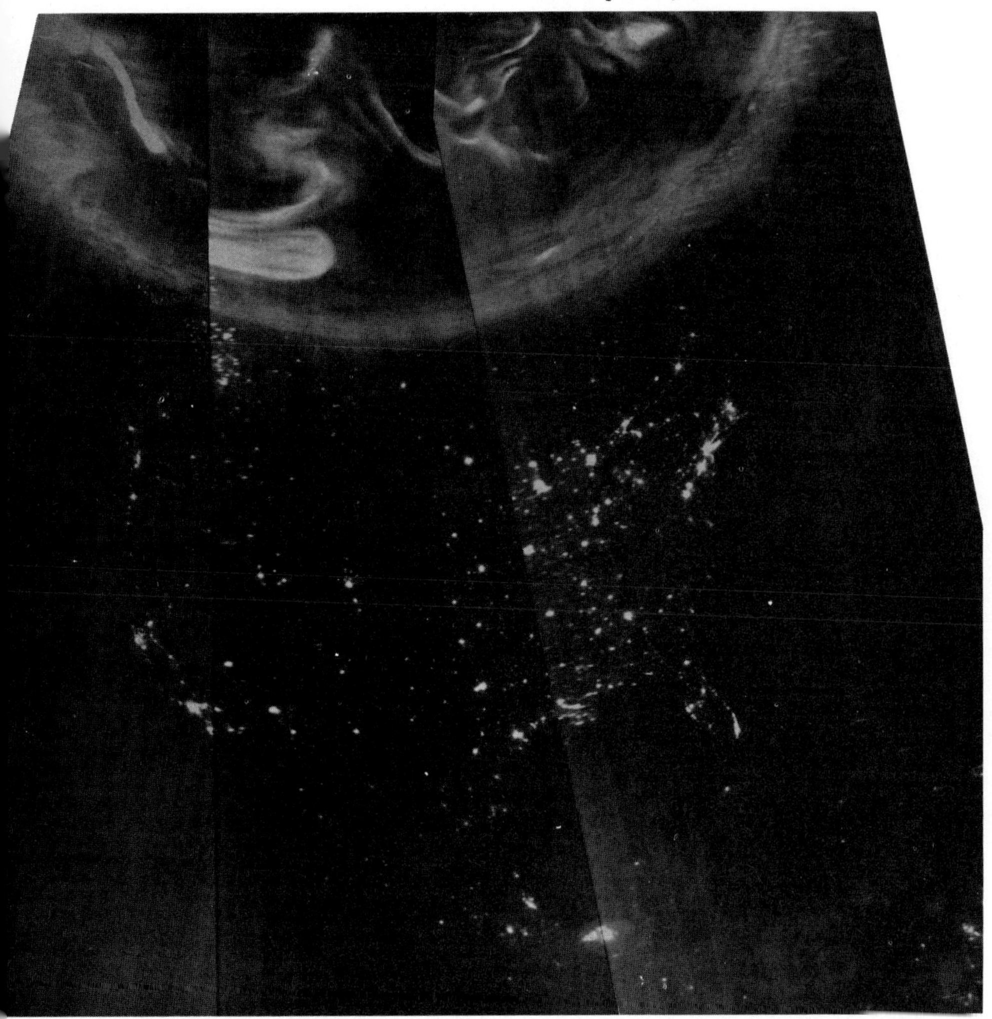

Figure 12-8. City lights and the aurora borealis, on a clear night over North America, photographed in the near infrared by the U.S. Air Force meteorological satellite DMSP-S, on October 19, 1974. Pictures from three successive orbits have been combined so as to show the entire continent from Atlantic to Pacific. However, there is some slight distortion at the edge of each picture. Changes in the structure of the aurora borealis during the 102 minutes between each photograph are apparent. (photographs courtesy T. A. Croft, SRI International).

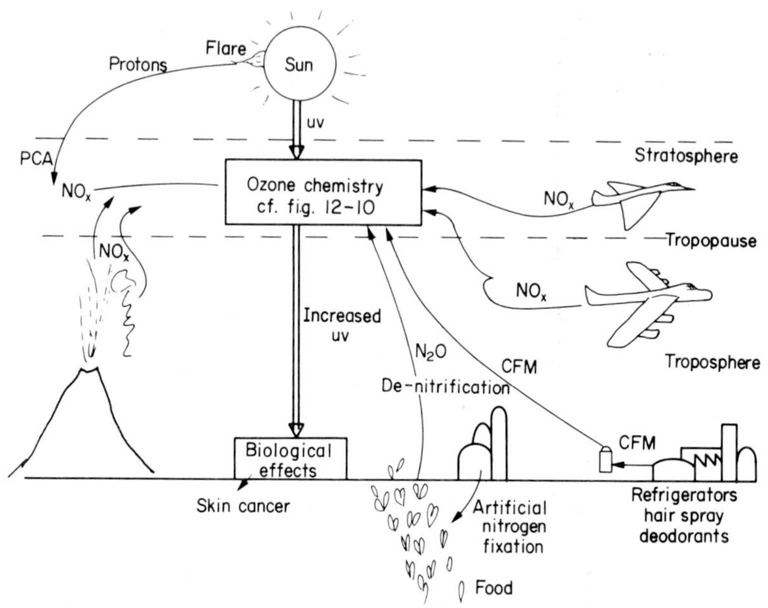

Figure 12-9. Natural and anthropogenic processes affecting the ozone content of the stratosphere.

above. It was estimated in 1977 that accumulated chlorofluoromethane production may already have led to a 3% depletion of stratospheric ozone, and that if world production continued to increase, this figure could reach 18% by the year 2000. Even if measures are rapidly taken (as in the United States and Sweden for chlorofluoromethane aerosol propellants) to restrict the use of these compounds, ozone depletion will continue for some decades as CFM's presently in the troposphere gradually reach the ozone layer. This illustrates the urgency of evaluating the environment impact of new technologies *before* they are applied in a major way.

Considering the effects of the nitrogen, chlorine and hydrogen reaction chains separately (see the boxes in Fig. 12-10), everything seems to work to destroy ozone. Many early estimates of the threat to the ozone layer were based on such simple evaluations. However, the interactions of these different reaction chains must be considered. At the top of Fig. 12-10 (lozenge A), we see that NO and ClO can react, thus bypassing part of the ozone destruction cycles. We see also (triangles B and C) that NO_2 can react with the hydroxyl radical HO to form HNO_3, and that Cl can react with HO_2 to form HCl, and these two acids have a good chance of being

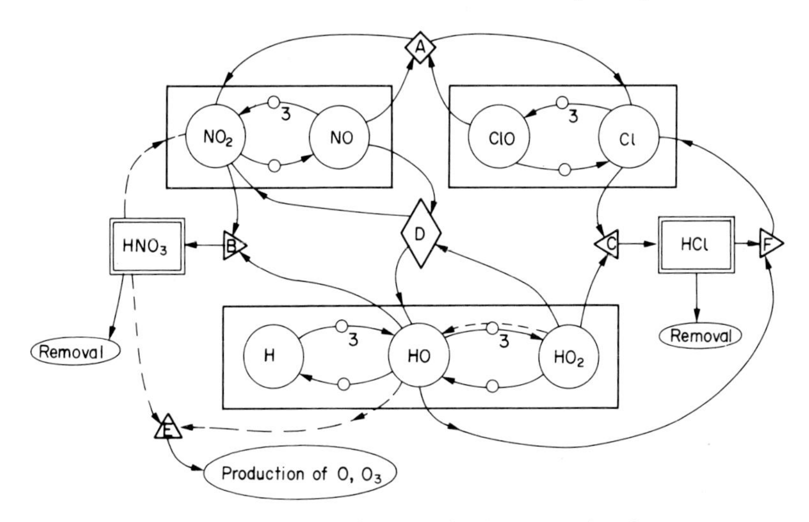

Figure 12-10. Aspects of stratospheric ozone chemistry.

removed from the stratosphere. This is a limiting factor on ozone destruction.

A new development in 1978 was the discovery that the reaction between HO_2 and NO (lozenge D) proceeds much more rapidly than previously thought.[10] The overall picture is changed as a result. At the altitudes (18 km) at which the Anglo-French Concorde flies, the net result of NO_x emission becomes a slight *increase* in the ozone content, because the reaction path from NO to D to NO_2 to B to HNO_3 combines at E with the reaction path HO_2 to D to HO to E, and produces more ozone than is destroyed in the nitrogen box. This is not the case above 20 km, where the American SST was planned to fly; there NO_x remains an ozone destroyer. While the new developments have somewhat mitigated the effects of the nitrogen oxides, they have aggravated those of the chlorofluoromethanes. With less NO surviving, the ClO to A to Cl bypass is weakened, and much of the extra HO produced by D reacts with HCl at F to produce additional Cl, thus intensifying the cycle of ozone destruction by chlorine. There may be more surprises in store for us, and Fig. 12-10, complicated as it may appear, remains a gross simplification of the real picture.

Present impact on the ozone layer of air transport and of CFM release may be slight, and it seems possible to limit future impacts without enormous difficulty. This may not be the case for the impact of human agricultural practices. As mentioned earlier, Man has intervened in the global nitrogen cycle in a massive way, substantially increasing the rate of nitrogen fixation by the application of industrial fertilizers. At present, about 50 million tons of fixed nitrogen are applied in this way, and there is strong pressure in favor of further growth of this figure, in order to keep up with world population growth. Denitrification is estimated to return about 7% of this to the atmosphere in the form of N_2O. Thus 1.4 million tons might be released in the year 2000. This gas can reach the stratosphere, and there be transformed into NO_x through reactions with photons, not shown on Fig. 12-10. Estimates of the rate at which ozone might be depleted vary widely, ranging from zero or even an ozone enrichment, to as much as 23% depletion by the year 2000. Most scientists believe that the matter is not so urgent, but in the long run it could become a serious problem, because it involves committing ourselves to causing a certain degree of harm in order to maintain or increase food production. It is not simply a matter of justifying the economic costs of finding replacements for chlorofluoromethanes, or of developing jet engines with low NO_x emission, against an additional skin cancer mortality of a few thousand per year (mostly privileged enthusiasts of sunbathing in the rich countries); it involves vital food production for many millions of people.

At the present time, the impact of the human species on the planet Earth is far from negligible, even though it is not overwhelming. Until very recently, the major impact was linked to the simplification of ecosystems and the transformation of the Earth's surface by the spread of agriculture. Over the last 10 000 years, and especially over the last few decades, we have significantly perturbed the natural cycles of carbon and nitrogen in the biosphere and atmosphere, adding totally new processes and enormously accelerating some of the existing ones. While absolute ecological equilibrium has never existed in the biosphere, the ultrarapid cultural and technological evolution of Man has led to a pace of change far more rapid than biological evolution, except for that of microbes and to some extent insects. The global ecosystem is now far from equilibrium, since our numbers are rapidly growing. Our inadvertent impact on the climate represents a new challenge to the feedback mechanisms of the geosystem, and we have only a vague idea of what the response may be. Insects and bacteria can adapt quickly to changing environmental conditions. The question for us is — Will the future environment of Man on Earth, *which we are now in part inadvertently shaping by our actions*, be one which *we* will find comfortable?

13

The Future of Humanity

Some say the world will end in fire,
Some say in ice.
From what I've tasted of desire
I hold with those who favor fire.
But if it had to perish twice,
I think I know enough of hate
To say that for destruction ice
Is also great
And would suffice.

Robert Frost[1]

THE future of the Earth as a planet is bound up with the future of the Sun as a star, and it is assured for the next few eons. Only when the supply of hydrogen in the core of the Sun becomes exhausted, and rapid evolution toward the red giant stage ensues, will conditions on Earth necessarily become inhospitable to the continuation of life. When the solar luminosity will have increased by a factor of five, the Earth's effective temperature will be above 100°C, and in all likelihood the oceans will have boiled away as a result of a runaway greenhouse well before that point is reached. No doubt the solid Earth will survive the red giant phase of the Sun, but the cinder that will remain orbiting a rapidly cooling white dwarf will not be a very interesting place. Thus the days of the planet Earth are numbered. Such a life expectancy may seem very long, but it is quite short compared to the life expectancies of the most common stars, the red dwarfs, which burn their hydrogen so slowly that they may continue in the main sequence phase for tens if not hundreds of eons. It is conceivable that inhabited planets may be orbiting such stars, somewhere in the universe.

Of course to anyone concerned with the history and evolution of human society, an eon is practically the same as infinity. Indeed, even the climatic history of the last 100 Myr, for the most part completely ice-free, is irrelevant to the cultural evolution of Man, although it determined the biological evolution leading up to *homo sapiens.* Man is a creature of the

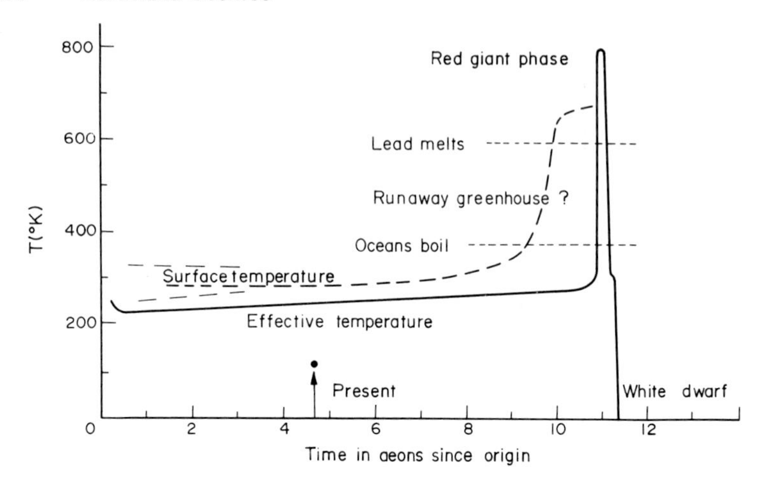

Figure 13-1. The very long term prospects for the planet Earth.

current Pleistocene ice age, and civilization is a phenomenon of the interglacial period which began about 13 000 years ago.[2] Over the next 100 000 years, it is extremely likely that the Earth will again go through one or more glacial periods, with ice covering large areas in North America and northern Eurasia, whether or not human activities have any effect on climate. There seems to be no reason why the human species could not survive such a glacial epoch. Industrial-agricultural civilization could perfectly well prosper in such a period, although not in the same places as now. Of course, the prospect that one or more sovereign states jealous of their independence may have their territories completely or partially covered by ice, is not one that politicians or even "statesmen" are ready to think about. And yet we cannot assume that such changes will necessarily be gradual and nearly imperceptible. Rapid advances of ice in less than a century are known to have occurred, notably in the "younger Dryas" period of 10 800 B.P. If some such advance were to occur in the next few centuries, the most serious consequences would include not only the migrations of peoples directly affected, which could be large judging from the present population densities of the northeastern United States and central Europe, regions which might well be covered; but also the displacement of the major climatic belts and thus of the principal zones of food production. On the other hand, large areas of new land would become available as the sea level fell. Adapting to such changes in less than a century would certainly test the organization of world society, but it need not destroy it.

This is a rather optimistic view regarding the adaptive capacity of human society in the future. In contrast, observation of present-day tendencies may lead one to be more pessimistic. Although the specter of nuclear war between the two superpowers has receded somewhat over the last few years, we continue to live under the sword of Damocles of enormous arsenals of nuclear weapons. These are part of the "balance of terror", and one can well imagine this equilibrium to be unstable. If these arsenals of destruction are ever used in a massive way, very large numbers of human beings will die immediately or within a short time thereafter, a major portion of the world's agricultural and industrial productive capacity will be destroyed, and large land areas rendered dangerous to humans for some time. There will be strong perturbation of the biosphere, and this together with the increased mutation rate due to continuing radioactivity will set in motion ecological transformations whose scale and nature are difficult to predict. The climate itself may be affected through various indirect effects. All this would probably not mean an end to human life, nor even a return to pre-industrial society, but it would certainly constitute a catastrophic failure of civilization.

Let us hope that such a failure can be avoided. Other extremely serious problems face humanity. In the preceding chapter we described how Man's extremely rapid cultural and social evolution has transformed the environment, mainly through the spread of agriculture. From the few million of pre-agricultural times, the human population rose to some 800 million in 1750. Most of this growth and development took place in Asia, and there the environment was the most radically transformed. Certainly in the 15th century, a division of the world into developed and developing countries would put Asia into the former category, and Europe and America into the latter. Even in 1750, human environmental impact was probably roughly proportional to population, with some enhancement where population densities were particularly high. The energy use *per capita* was probably fairly uniform, except for heating needs where winters were cold.

This situation has changed enormously over the last two hundred years. Between 1750 and 1920, the industrial revolution (and the opening of the Americas to agriculture) led to a nearly sixfold increase in the European population (including emigrants to the Americas, and their descendants). Although Asia's population doubled during that same time, its share of the world population fell from 80% to roughly 60%. During this time, *per capita* energy use was rising much more rapidly in the "West" than

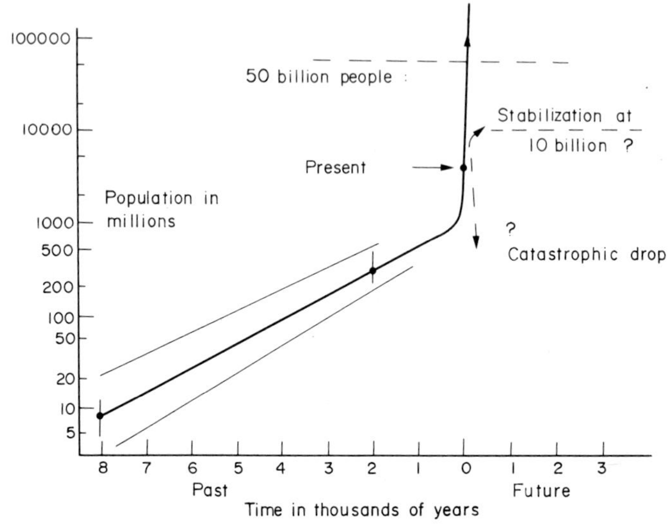

Figure 13-2. World human population growth over the past few thousand years.

in Asia, so that the environmental impact of human activities became about the same in these two areas.

Over the past fifty years, the continuation of some of these trends, together with the appearance of new trends, have led to a radical division of human society, between the industrially developed countries of the "North", and the poor industrially underdeveloped countries of the "South". In these Third World countries, the introduction of effective death control has led to a rapid rise in the rate of population increase, so that since 1920 the population of Asia has more than doubled, and its share of the world population is once again growing and will soon reach 80% as in 1750. The environmental impact of the increased food production is enormous: in Asia, practically all cultivable land is now under cultivation; in Africa and South America, the proportion under cultivation is increasing rapidly.

In the North, the major phenomenon has been an enormous rise in the *per capita* energy production, coupled to a rate of population growth much lower than that in the South. This "demographic transition" is the result of the adoption of birth control methods as effective as the death control techniques already introduced. It accompanies the rise in *per capita* energy consumption to a level ten times that in the South, which is fairly well correlated with Gross National Product *per capita.* Many

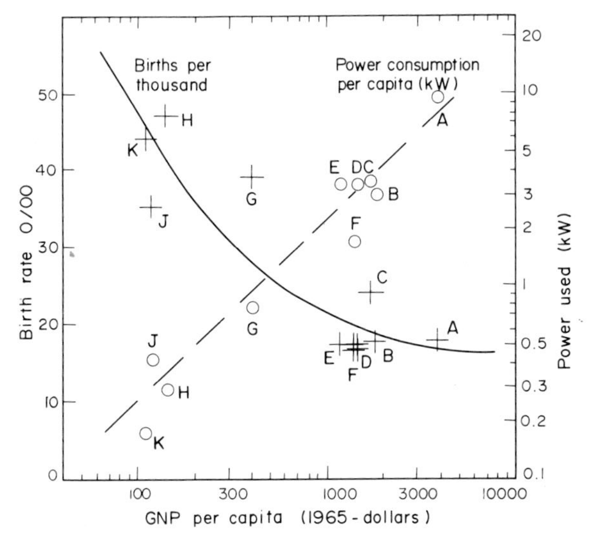

Figure 13-3. Birth rates, power consumption and *per capita* GNP.[3] Regions are: A — US and Canada, B — Western Europe, C — Oceania, D — USSR, E — Eastern Europe, F — Japan, G — Latin America, H — Africa, J — Communist Asia, K — Other Asia. In the developed, rich countries, the birth rate is low. In the (hopefully) developing, poor countries, which constitute three fourths of the human population, birth rates are high. Is high power consumption a necessity for, or a sign of wealth?

demographers believe this to be no mere coincidence, and maintain that effective birth control depends first of all on the elimination of poverty. At any rate, the increased Gross Northern Product is coupled to an increased Gross Northern Pollution, and probably the industries of the Northern countries are responsible for much of the rise in atmospheric CO_2. At the same time, the needs of the industries of the North for fuel and raw materials have grown enormously, and many of the easily accessible sources of such materials in the North have already been exhausted. The exploitation of other northern resources, such as the coal under the western Great Plains of America, or oil and coal in the Siberian north, is very expensive. Such exploitation has hardly begun, but raw materials for American and European industries are being extracted all over the globe. Despite their limited population growth, the Northern countries' impact on the global environment is growing rapidly.

It is clear that the current trends cannot continue for very long. The world population has already passed four billion, and following current trends it will pass six billion before the year 2000, with close to five billion in the less developed countries of the South. It is not easy to feed such a population, but it certainly is possible, combining modernization of agriculture with an expansion of cultivation in areas of South America and Africa where land is still available. Indeed it has been suggested that the Earth can support as many as fifty billion people. That may seem like a lot, but at a growth rate of 2% per annum that figure would be reached in less than 130 years. If birth control does not succeed, death control will certainly fail.

The present world human society appears to be one-fourth "rich" and three-fourths poor, with the rich getting richer, and the poor getting much more numerous. Moreover, the wealth of the rich seems to require exploitation of resources located in the lands of the poor. To a visitor from another planet, it would not be evident that we are all one species, despite the nearly unanimous pronouncements by our leaders since 1945 that all men are brothers. It would be perfectly clear, by contrast, that the continuation of the present trends must lead to catastrophe, i.e. to a sudden change. While one hardly needs to mobilize a giant computer system to see this point, it has been dramatically illustrated by the various computer models of the Club of Rome report, many of which predict a breakdown of industrial world society, survived by a much reduced population in a world despoiled of easily accessible resources (except for scrap).

Most people hope of course that catastrophe is not inevitable, i.e. that there can be a gradual (but necessarily rapid) adjustment to a more stable situation with a relatively constant world population, obtained by a fall in the birth rate rather than a rise in the death rate. The need for some reason to hope, an uneasy conscience about being well-off in a world where most people are poor, the theory of the demographic transition, all these factors combine in the doctrine that the solution to the present predicament of world society lies in rapid agricultural modernization and industrial development in the Third World, with improving living standards there leading to a fall in the birth rate. A diffuse egalitarian sentiment, seldom translated into effective action, promotes the idea that ultimately *per capita* consumption should be roughly the same everywhere. Since virtually no one will volunteer to have his present standard of living reduced, it is implicitly agreed that all countries should be brought up to

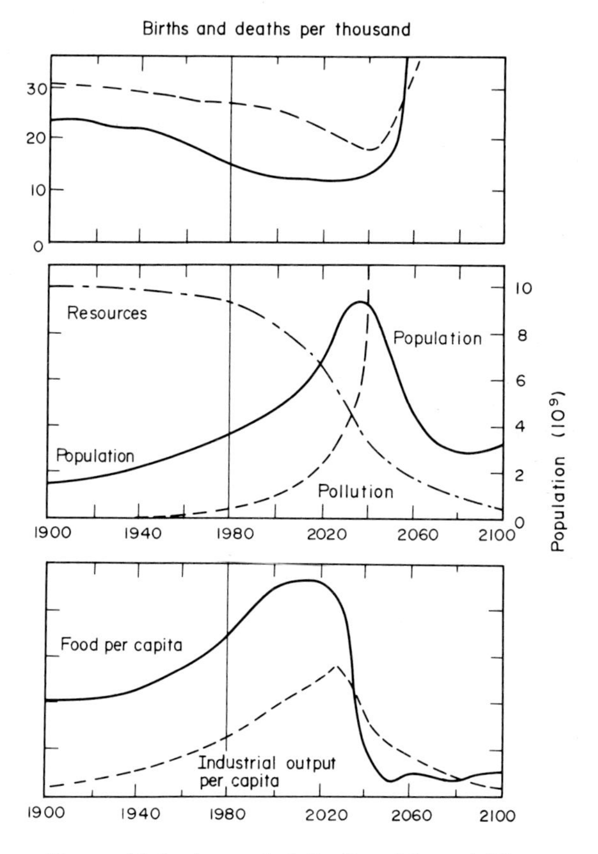

Figure 13-4. A pessimistic "world model" from the first Club of Rome report. Scales for resources, pollution, food and industrial output *per capita* are arbitrary. The basic assumption is that the world is a single unified system, i.e. global averages are used and regional differences neglected.

the American standard. Since even Americans will not admit that they are well enough off (and not all of them are), but would like to be still better off, it is accepted that the entire world should be aligned with a U.S. *per capita* consumption rather higher than today's. Assuming that a demographic transition can be achieved without constraint by sufficient economic development, this means that we must move fast if things are to remain at all manageable. Given the momentum of population growth, it seems certain that the world population will reach at the very least eight billion, barring an earlier catastrophic decrease. Present U.S. energy consumption is about 10 000 watts *per capita*, as contrasted to average world consumption of 2000 watts *per capita.* If by eliminating waste, of which there is a great abundance, the U.S. standard of living can be improved with no further increase in *per capita* energy consumption, then to supply eight billion people with ten kilowatts each means that we must reach a global energy production of $80\ 10^{12}$ watts (80 terawatts = 80 TW), about ten times the present level, some time in the next several decades.

This is still small (less than 0.1%) compared to the average solar input, and the direct global effect of this energy release may be negligible. However, the local and even regional thermal pollution where the energy is produced and where it is consumed may not be negligible, and there may be indirect effects on a global scale. The nature of these indirect effects depends for the most part on the strategy adopted for producing this energy. At present, virtually all energy production involves the combustion of fossil fuels, coal, oil and gas in roughly equal (energy equivalent) amounts. The share of hydroelectric and nuclear power is very small, that of geothermal or solar electric power smaller still. Note that about 5% of the present human energy consumption is solar energy stored in food and released by our metabolism. Now we know that our recent and present burning of fossil fuels has already noticeably increased the carbon dioxide content of the atmosphere, with a possible impact on the mechanisms determining the climate. What of the future?

Almost all strategies proposed for the development of global energy production foresee a continued increase in the consumption of fossil fuels, especially oil, until at least the year 2000. This is because it appears difficult to introduce new technologies (nuclear, solar) on a massive scale very quickly. Of course it implies a certain optimism, not necessarily justified, regarding the continued availability of such fuels at prices that will allow expansion of their use. Thus all these energy strategies commit us to an increase in the rate of CO_2 emission, at least until the year 2000. It seems very likely that the concentration of CO_2 will rise from the present level of 330 ppm to a level of nearly 400 ppm at the end of this century, corresponding, according to simple climate models, to a temperature increase of 0.6°C. This is a small but not negligible change, which may trigger further increases through positive feedback from the oceans.

A variety of different strategies are proposed for the 21st century, at least by those few governments, institutions and individuals able to look

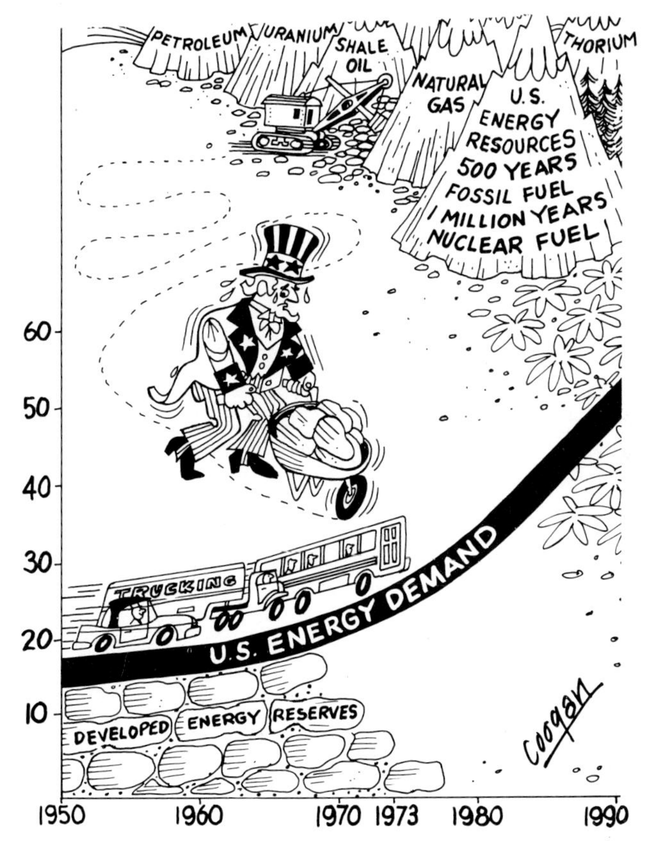

Figure 13-5. Even in the United States, with an energy consumption of 10 kilowatts *per capita,* the expectation is that energy use will continue to increase. Is such growth sustainable or desirable? From *Physics Today,* December 1973. © 1973 American Institute of Physics.

beyond a time horizon of twenty years. It seems clear that unless consumption drops sooner, oil and gas will run out early in the next century.[4] It will appear scandalous to humans of future generations, that we should have squandered these valuable nonrenewable resources, precious for all sorts of chemical applications, simply burning them. At any rate, for the coming century, many planners believe that the only solution is nuclear power, others favor a mix of nuclear and solar energy. Some planners believe solar energy to be impractical on a large scale, and nuclear power to be both uneconomical and too dangerous, and they note that enormous coal reserves remain untapped. An energy strategy based on coal will however lead to much more emission of carbon dioxide. Even without going above a global energy production of 50 TW in the year 2100, a coal-based strategy would lead to an atmospheric CO_2 concentration of 1500 ppm in the year 2100, i.e. over five times the pre-industrial level, and this would seem to entail a global average temperature increase of at least 8°C. Such an increase is very substantial, and it would probably trigger major changes in the global circulation and in the climatic belts, perhaps even leading to the melting of the polar caps.[5] Although this last would be a very slow process, it might be irreversible. The problem is that we do not yet know what all the consequences of burning coal to produce 50 TW might be, and yet decisions on energy research and development strategies must be made now. Redeployment of resources is neither easy nor rapid, where energy production systems requiring huge amounts of capital and long lead times for implementation are concerned. Perhaps a warmer world would be pleasant, at least for those people whose land stayed above water and did not become desert. However, the risks of a coal-based strategy seem great, and even the coal will not last forever.

Burning coal may have other disadvantages. At present, the combustion of coal contributes significantly to the global aerosol load. It has been suggested that by increasing the global albedo, such aerosols may have a cooling effect which can partially compensate for the heating effect of the CO_2 rise. However, recent work suggests that the net climatic effect of such particles, taking into account their optical properties and the fact that they are released and found mostly over land, is a net *warming* and not cooling effect. Such aerosols would only make things worse — they constitute a serious health hazard, and if coal burning were again to become a dominant energy source in the future, appropriate anti-pollution measures could, and we hope would be taken. However, the CO_2 emissions probably cannot be stopped without cancelling most of the energy benefit of burning coal.

In addition to the environmental hazards of *burning* coal, we would have to consider the impact of *mining* it in the huge quantities required for a coal-based energy strategy. These impacts would certainly be important on a local and possibly on a regional scale, possibly even on a global scale if so critical a region as Antarctica were to be exploited. What are the *alternatives* to a fossil fuel strategy for the 21st century? Until recently it appeared that most governments of the industrially developed countries,

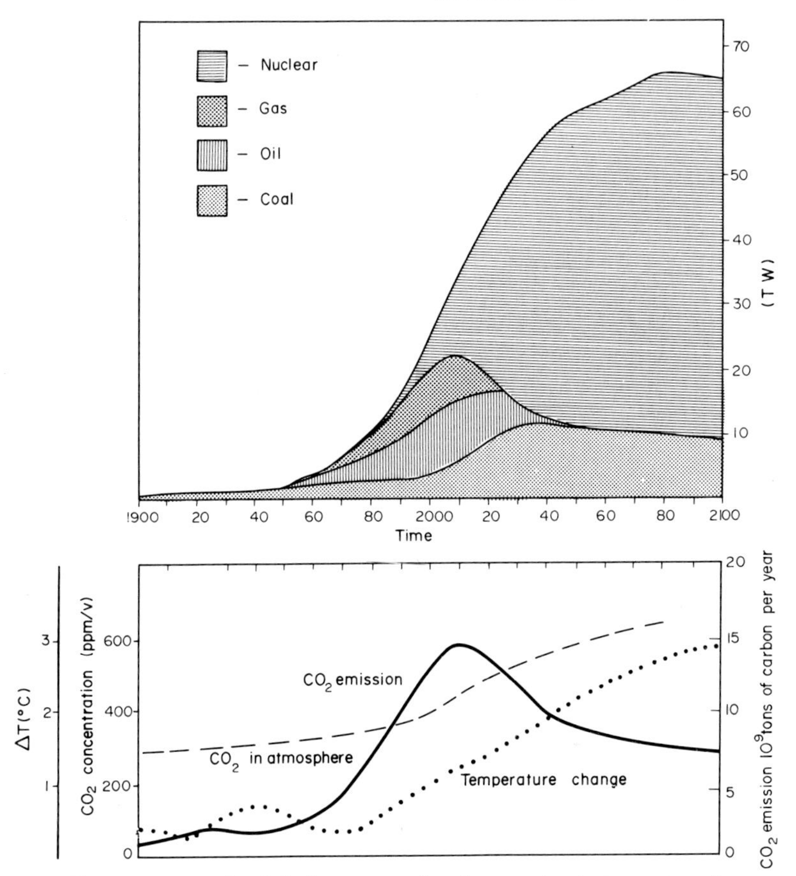

Figure 13-6. A global energy development strategy based ultimately on nuclear power (50 TW) still must make use of fossil fuels over the next few decades. Here we see that the impact of such a strategy on climate is not negligible, even after fossil fuel burning begins to decline. (Source — Niehaus & Williams, IIASA & IAEA).

and indeed also in the Third World, were counting on nuclear fission reactors, if they were at all thinking about the problem. Such reactors are slightly less efficient than fossil fuel power plants in generating electricity, i.e. more waste heat is usually released. However, the difference is not very large. In routine operation, in the absence of any incidents or accidents, sabotage, etc., there is little doubt that such nuclear power

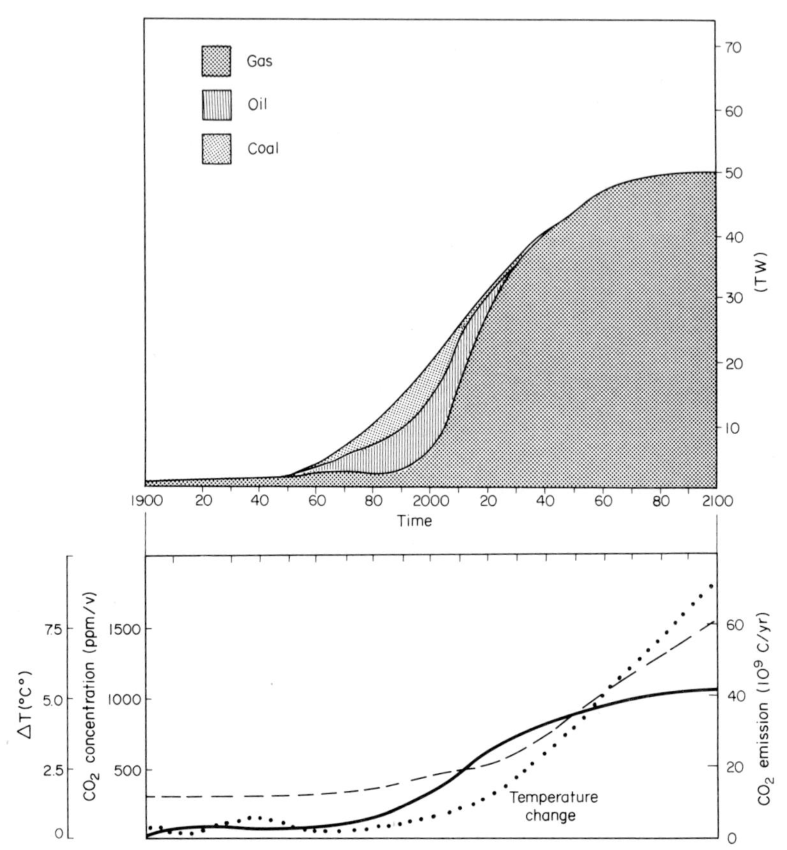

Figure 13-7. A fossil-fuel (mostly coal) based global energy strategy (50 TW) and its impact. Note that the scales in the lower graph are not the same as in Fig. 13-6. The predicted temperature change of nearly 10°C is enormous. (Source — Niehaus & Williams, IIASA & IAEA).

plants have less impact on the global environment than do fossil fuel power plants. In theory, very little gas or radioactivity should escape the reactor. By comparison, burning coal may well disperse *more* radioactive isotopes in the global atmospheric environment. There is of course the problem of reprocessing or otherwise disposing of the highly radioactive wastes, but here too it is claimed that if all goes as planned, disposal can be perfectly safe. Salt beds are known in which liquid water cannot have circulated for many thousands of years, and this is only one of a number of types of disposal sites said to have sufficient stability. It remains to evaluate the risks involved with some confidence. It is true that the waste products of the ancient natural fission reactor discovered in Oklo (Gabon, Africa)[6] are still there after 2 billion years, but one example hardly constitutes grounds for complacency. It has been suggested that by "burning" uranium in reactors we are cleansing the Earth of radioactivity. In a sense this is true, but creating serious local hazards for the next several thousands of years, in exchange for a diffuse global background for the next few eons, may not be everyone's idea of a good bargain.

The great controversy over nuclear power is not so much over the hazards of routine operation, as over the risks of a serious accident. In any serious cost-benefit analysis of a massive program of any sort, the extremely damaging even though highly improbable accidents must be taken into account: experience with dam failures, as well as the collision of two fully-loaded jumbo jets on Tenerife teaches us that. In fact, some evidence is finally emerging, after nearly 20 years, that such a highly damaging nuclear accident did indeed occur, near Sverdlovsk in the U.S.S.R., toward the end of the 1950's.[7] No doubt techniques have advanced significantly since then, but the problem remains that full-scale tests of some of the emergency systems of nuclear reactors are not feasible. There are still many other important issues connected with nuclear energy — political, economic, social. Governments have generally failed to obtain the confidence of their people that they are being completely candid regarding the risks involved. Considerable opposition has developed, and some governments are showing signs of hesitation. In any event, nuclear fission fuel supplies are finite, even if large.

In addition to pushing the development of nuclear fission power, several governments have been funding research in controlled fusion power. (Explosive fusion power is already available, in the hydrogen bomb.) Fusion, after all, is what makes the Sun shine. In the Sun this is made possible by compression due to the mutual gravitational attraction of some 300 000 Earth masses of hydrogen and helium, which keeps the central core of the Sun hot and dense. To keep a plasma hot enough and dense enough for a long enough time for fusion to take place and yield useful energy, at the surface of the Earth, is much more difficult. We need to use the electromagnetic force in some way for this; various devices have been made, such as the Tokamak pioneered in the U.S.S.R., and many experiments conducted. There seems to be progress, but we cannot say whether fusion power will be a significant contributor to human energy production before the end of the 21st century, and it would be

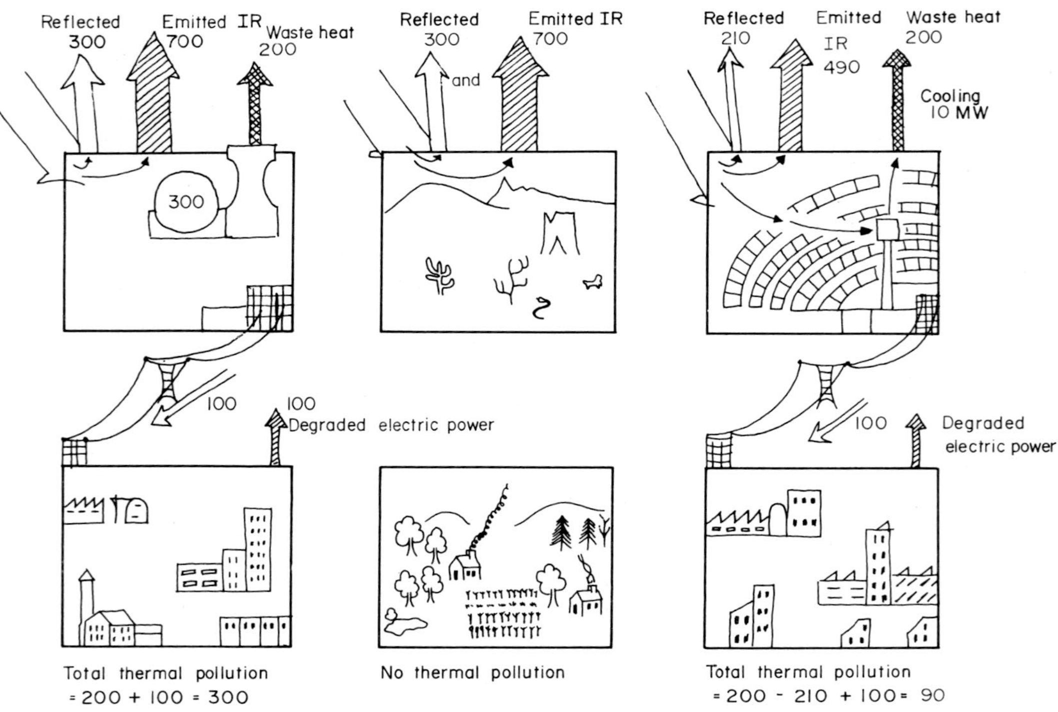

Figure 13-8. Power production and thermal pollution. The desert site receives an average of 1000 MW of solar power.

Left: nuclear plant in the desert site, supplying 100 MW of electric power to the city.
Right: the same power supplied by a solar facility.
Center: virgin desert; instead of the city, a small primitive agricultural settlement.

foolish to count on it. Since fusion is based on hydrogen (actually the moderately rare deuterium isotope $_1H^2$ would probably be used), ordinary water can be the source of the fuel. Since the energy yield is fantastically high (over 3000 kilowatt-hours per kilogram of water), the energy resources available would be enormous, inexhaustible over many eons. A nuclear fusion power plant would be a potentially dangerous place, since very-high-energy nuclear processes are involved, but in general, the global environmental problems associated with fusion power are likely to be less serious than with fission. Thermal pollution limits, at the place of consumption, will of course apply to fusion power if it comes to pass. Even if fusion power could supply 10 000 TW, such a level would be intolerable because it would make the Earth too hot a place for us to be comfortable in.

There is one energy resource that will be available as long as the Earth is habitable, and that is the Sun. We have suggested that 80 TW of power could support a world population of eight billion quite comfortably, and yet 80 TW is less than 0.1% of the total solar input to the Earth. This might require consecrating an area of 800 000 square kilometers to gathering solar energy, e.g. 8000 solar parks of dimensions 10 km × 10 km. Such areas are very *small* compared to the areas devoted to agriculture (another form of solar energy production). Tropical lands would be favored, but the distribution of energy would still be far more egalitarian than the present distribution of fossil and nuclear fuels. Moreover, the global circulation of the atmosphere and oceans transfers large amounts of (originally) solar energy north and south, where some of it can be used as wind or wave power, or tapped by exploiting thermal gradients in the ocean. There are many technological problems associated with economic exploitation of solar energy on a large scale, but they do not really appear to me to be more difficult than the problems associated with nuclear power, especially if one is willing to consider different patterns of development, less urbanized or centralized than in the presently developed industrial countries of the North. This may be a distinct advantage where development of the (mostly tropical) Third World countries is involved. In my opinion, if the future of humanity is to be bright, it must be solar.

This is not to say that the environmental impact of solar energy exploitation is negligible. The problem has hardly been investigated. Certainly the impact on the land covered by a solar power park can be significant. Lovers of desert landscapes may not be pleased to see a field of thousands of mirrors where once there was only sagebrush and cactus. Still, the insult to the environment will be far smaller than that of coal mining operations and the Four Corners power plants. The air will be unaffected, except to the extent that wind patterns are changed. On a global scale, the thermal impact will be much smaller than that of either fossil fuel or nuclear (fission or fusion) power plants. The reason is that a solar power park will generally not absorb much more solar energy than would in any case have been absorbed at the site. Even when the natural surface albedo is as high as 30%, reducing it effectively to zero makes little difference. This is shown in Fig. 13-8 where we compare the impact of a solar facility producing 100 megawatts of electric power, with that of a nuclear reactor producing the same amount. Note that this is big for a solar power plant, but small for a nuclear one, by present technological standards. In each case we assume that the site is 4 square kilometers of desert, with an albedo of 30% and average insolation (over 24 hours) of 250 W/m^2, and that 200 MW of waste heat are produced and released locally. This implies that 300 MW of solar power are being fed into the generating station itself, and this corresponds to 1.2 km^2 of effective sunlight collecting area with zero albedo as far as reflection to space is concerned. Thus it is reasonable to assume that the solar power facility absorbs only 90 MW more than the virgin desert, and since it "exports" 100 MW of electric power, there may be a slight local cooling effect. Of course, nearly all of this electric power ends up as heat at the point of use, so 90 MW are indeed added to the geosystem. By contrast, the nuclear power plant adds 300 MW of heat to the geosystem, this thermal pollution being divided between the point of production and the point of use. If the area in which this power is used is 10 km^2, the thermal pollution resulting from the degradation of electric power might amount to over 5% of the natural solar input there. Of course, this will be true for any imported energy, whatever its source. However, from a global point of view, thermal pollution by nuclear power (or fossil fuels) is at least twice as large as when solar power is used. Even if nuclear fusion power turns out to be feasible and very efficient, it almost certainly involves much more thermal pollution than solar power production. If solar power is used, even a global power production of hundreds of terawatts can in principle be contemplated; there would be many local problems, and possibly regional ones, depending on the pattern of energy production and consumption, but global thermal pollution would remain small.

Man does not live by energy alone; food is even more essential. The global environmental impact of growing food for a population of eight billion is enormous. High-yield agriculture of the sort that has given rise to the term "Green Revolution" involves large-scale application of industrial fertilizers as well as the use of selected strains of wheat, rice, etc. We explained in the preceding chapter how the use of artificial fertilizers perturbs the nitrogen cycle of the Earth, with consequences in the stratosphere as well as on the ground and in the sea. We do not really know what further massive modification of the nitrogen cycle might do. As for the use of only a few selected strains, many ecologists have warned that such a practice — a further radical simplification of the biosphere — continues the trend which began with the invention of agriculture several thousand years ago, and which is not without danger. The risk is that such simplification renders us particularly vulnerable to insect or microbial pests which may specialize in living off our selected strains, and do this even more effectively than we. An example is the Irish potato blight and famine of the last century. The use of pesticides such as DDT may bring temporary relief, but it creates further problems for us as insect pests rapidly develop resistance through biological evolution by natural selection, while we accumulate DDT in our tissues.

The farming of the sea has often been suggested as a solution to our food problems. However, there is strong evidence that we are already close to overexploiting the resources of the sea, except possibly where the "krill" of Antarctic waters are concerned. Marine life is limited by the need for nutrients, and these are brought up from the sediments of the sea bottom by upwelling. It has been suggested that such upwelling could be artificially enhanced by placing nuclear reactors on the sea floor, with the waste heat release driving convection. At present, one can certainly be unenthusiastic about the idea, because of the risks of contamination of marine life by radioactivity. For the more distant future, it remains a possibility, but effects on ocean currents and on ocean—atmosphere heat exchange will have to be evaluated. It should be noted that many developed countries with seashores, such as the United States and France, already count on using the ocean as a sink for the waste heat released by their planned nuclear reactors, without too much worry about the impact of this. Some solar power concepts also involve the ocean, and if developed on a large scale will modify ocean currents and water temperatures locally. It may well be possible to use the oceans both for food production and in connection with energy production, and even for disposal of various wastes, but careful planning will be necessary to ensure that the various strategies of use are compatible with one another. In any event, continued biological productivity of the marine environment requires that present trends of pollution by oil, DDT and various other materials *not* continue.

The arguments given here are conservative, in that they assume that radical changes in the life styles of inhabitants of the developed countries may not be necessary and they are optimistic, in that they assume that the North-South dialogue will lead to transfers of technology and of wealth to the Third World, that energy consumption *per capita* will grow very rapidly there, and that population growth will decelerate and be halted. This is extremely urgent. A world of 8 billion people consuming 80 TW in the year 2100 is difficult but perhaps possible to attain, and it could be liveable. By contrast, if the demographic transition in the Third World is delayed, the problem may be one of a world of 15 billion people desiring to consume 200 TW in the year 2100, which would involve far greater environmental modification, and would not be nearly so liveable, even if it could be organized, which is far from certain. Science fiction writers and technocrats may be able to imagine hyperorganized societies of several tens of billions of people consuming thousands of terawatts, but few of us would want to live in them. There are not many generations between us and the year 2100, and it seems doubtful whether our great-grandchildren would be happy there either. Even today pre-industrial attitudes make adjustment to our present social organization difficult.

I do not believe that a world society one-quarter rich and three-quarters poor will survive for long. Even if nuclear war can be avoided, the continuation of such a pattern implies not just a moral blindness, but also an unwillingness to face the constraints imposed by a finite Earth. If the rich nations are permitted to allow the poor peoples of the world to starve, they are likely neither to take the measures necessary to prevent catastrophic modification of the environment by their own activities, nor to adjust to and plan for the depletion of easily accessible resources. In addition to catastrophic famines in the poor countries, there will be local and then regional environmental disasters and collapse of industrially overdeveloped societies. Such behavior is perfectly natural: many species have evolved in catastrophic ways and disappeared. For Man, the question is not one of biological but of social and cultural evolution. Our intelligence makes such rapid evolution possible, and it also allows us to evaluate, to some extent, our own behavior, and to try to predict where it is taking us. The question is whether we can translate this awareness into an effective feedback mechanism, controlling our actions so that there will be no such thing as *inadvertent* climate modification, or exhaustion of an essential irreplaceable resource, for example.

I do not agree with those who would propose a religion of the "natural" environment; there is no way back to pre-industrial days, and if present-day industrial and agricultural society collapses completely, the few million surviving humans will still be living in an unnatural environment showing the mark of that society, nearly everywhere on the globe. But it seems clear to me that while we are not yet in a position to engineer our environment, and unlikely to agree on what sort of engineering would be desirable if it were possible, we are in a position to understand some of the consequences of our actions on the global environment, and to avoid blind tinkering. The passage of Genesis (1.28) — "Be fruitful, and multiply, and replenish the earth, and subdue it; and have dominion over the fish of the sea, and over the fowl of the air, and over every living thing that creeps upon the earth" has often been criticized in recent years, as being contrary to an environmental ethic. However, as the great Jewish philosopher Maimonides put it 800 years ago, — "Here it is not meant to say that man was created for this purpose, but only that this was the nature which God gave man."[8] This is indeed the nature of any living species; not all have succeeded. When other species have failed, it has been blindly; but we can see where we are going, even if imperfectly. Need humanity fail?

In his *Modest Proposal* written in 1729, Jonathan Swift attempted to rouse the Irish people with a suggestion that "one fourth part of the infants under two years old be forthwith fattened, brought to market, and sold for food, reasoning that they will be dainty bits for landlords, who, as they have already devoured most of the parents, seem to have best right to eat up the children." The proposal was not adopted. No similar proposal has been made for Third World children now. A century after Swift, during the Irish potato famine, the solution for many was emigration to America. Today we sometimes hear suggestions that the solution to the population problem on Earth may lie in emigration to space. This is no solution. We have seen that the Moon and the other planets of the solar system are extremely inhospitable to life, and while one can easily imagine large scientific stations on the Moon and on Mars in the 21st century, colonization by large numbers of people seems out of the question. The exportation from Earth of some 100 million persons per year is not reasonable (for one thing, who would volunteer?), and yet this is what would be needed at current rates of population growth.

Space will no doubt play an important role in the future development of humanity, again, barring early catastrophe. One proposal, pioneered by Peter Glaser in the United States, involves the construction of satellite solar power stations in geostationary orbits around the Earth, solar energy being converted into a microwave beam for transmission to receiving stations on the Earth's surface.[9] The advantages include nearly continual sunshine, unabsorbed by air, at such stations, as well as an easing of structural problems given the weightless conditions. Rather detailed studies of the feasibility of such stations have been made. The economics depends on the cost of the materials for converting sunlight into electricity, the cost of transferring the materials into orbit and constructing the station there,

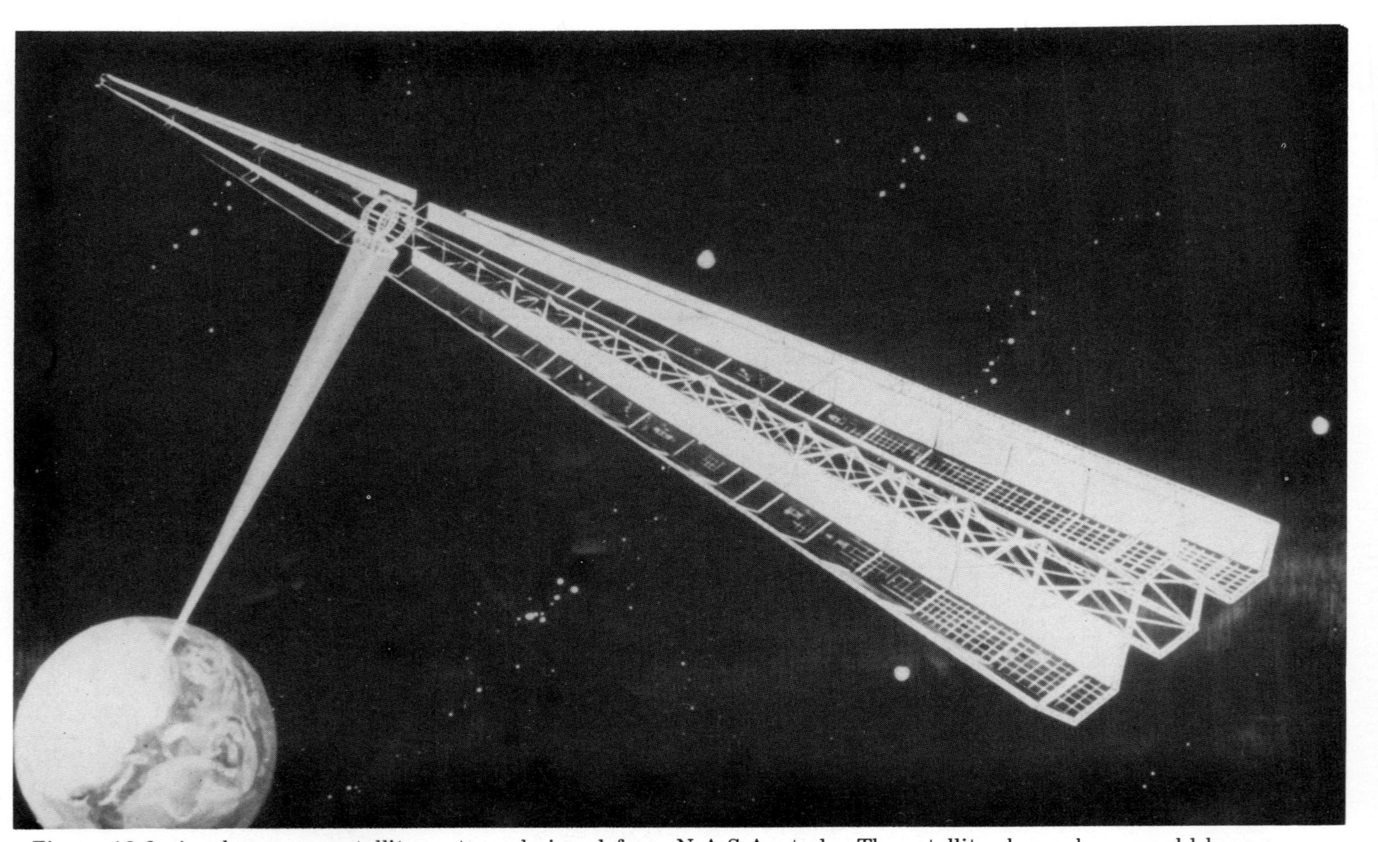

Figure 13-9. A solar power satellite system, designed for a N.A.S.A. study. The satellite shown here would have a mass of 50 000 tons and would transmit 5 gigawatts of electric power to a receiving station on Earth. Photo courtesy P. E. Glaser, Arthur D. Little, Inc.

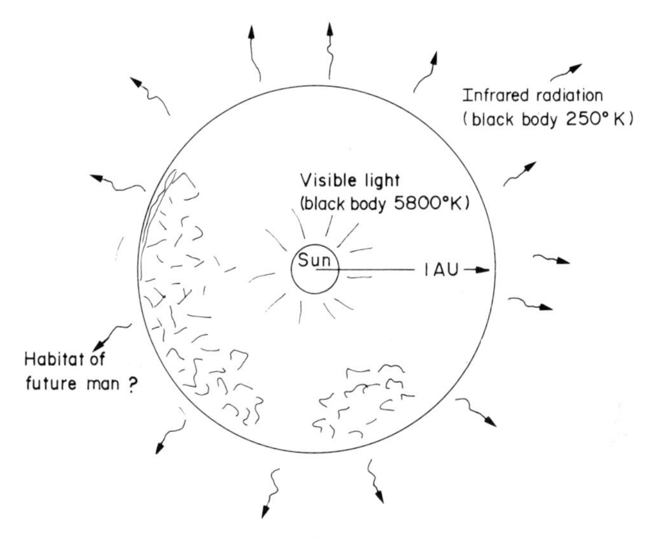

Figure 13-10. The ultimate conquest of space? Engineering of the solar system, by construction of a "Dyson sphere", so as to harness the entire energy output of the Sun for the purposes of an advanced technological society. Fantasy, dream or nightmare?

and the expected lifetime of station components. The plans are based on the NASA space shuttle, which will shortly be available, but there is controversy regarding the true costs of shuttle operations. The construction of such a station would require a large number of shuttle flights. Major rocket launchings, such as Skylab, are known to have perturbed the ionosphere,[10] and the effects of continual shuttle operations, needed if space solar power is to become significant, might be quite substantial. Apart from this, it must be noted that space solar power represents a net increase in the solar input to the Earth's climate system, much more than for ground-based solar power. Thus thermal pollution limits apply. Besides these environmental impacts of routine operation, there are the hazards associated with possible failure of control of the microwave beam transmitting the converted solar power to Earth. This is discussed in the project, but it should be noted that there is disagreement as to the maximum safe level of microwave irradiation for humans, Soviet safety standards being much more severe than American.

There are other uses of space which may be more important. Applications to telecommunications and Earth monitoring are well under way. Applications to certain industrial processes are already under study. For certain operations such as growing large perfect crystals, weightless conditions may be a distinct advantage. Although large amounts of material or energy may not be involved, such activities could become extremely important to future industrial technology. In the long run, many other industrial activities might be carried out in space. Extraction of mineral resources from the Moon or even asteriods has been proposed as an economic affair in the not too distant future, and would have the possible advantage of not polluting the environment of Man on Earth. Similarly, many industrial processes involve thermal pollution, and carrying these on in space using solar power would allow an increased level of activity without affecting the Earth's climate. All of this assumes that the impact of the rocket flights themselves will be small, which seems unlikely.

In the long run, economically self-supporting communities might be established in space, in small artificial worlds orbiting the Earth (or the Sun), using materials from the Moon or asteroids. These would certainly be no solution to the present world population problem, but they might provide opportunities for future expansion, and for a variety of life-styles. This begins to be the domain of science fiction, but such ideas are entertained by serious scientists, such as Gerard O'Neill at Princeton.[11] If we overcome the problems of the present and coming centuries, humanity may well be able to expand through the solar system. On very long time scales, the transformation of environments such as those of Mars and Venus is not inconceivable. Alternatively, one might imagine humanity using the material of asteroids to construct a very large number of artificial worlds intercepting a large fraction of the total output of the Sun. The extreme would be to surround the Sun with a sphere where *all* of the solar output is intercepted and exploited and ultimately converted into thermal infrared radiation, multiplying Man's living surface by a factor of 500 million. From afar, an astronomer would no longer see the Sun; instead he or she would see a "Dyson sphere", a large object, perhaps two astronomical units in diameter, radiating in the infrared, with an effective temperature of 250–280°K.

In this book we have looked into the various connections between our environment on Earth and the cosmic environment. We began by showing how the structure of the universe as a whole determines the shape of things on Earth, how the history of the Galaxy is our own history. We continued by examining in detail how the Sun rules the Earth, and how various terrestrial processes are governed by cosmic rhythms, even as the geosystem shapes its own response. We noted how the environment of the Earth has been transformed by the phenomenon of life, and how the spectacular success of the human species is itself modifying the globe, approaching limits which will impose an end to current trends quite soon. Our radio and television signals tell the distant astronomer that we are here. Will the tale be cut short? Or will humanity be able to emerge into a galactic future? That depends on whether we can control our own actions and shape our future in accord with the terrestrial and cosmic forces that govern us.

Appendices

APPENDIX I. MATHEMATICAL NOTATION, PHYSICAL TERMINOLOGY, AND UNITS

Mathematical notation

THE mathematics in this book is at a very elementary level. The notions of addition, subtraction, multiplication and division are surely familiar to all, but in addition to these operations, we shall frequently have use for the operation of *exponentiation.* To begin, let us consider an arbitrary nonzero quantity Q. We shall define exponentiation by writing that a product in which Q alone appears as a factor, n times, is equal to Q^n, which can be read as *Q exponent n,* or as *Q to the nth power.* Thus,

$Q \times Q = Q^2$ often read as Q squared,
$Q \times Q \times Q = Q^3$ often read as Q cubed,

and

$Q \times Q \times$ (n times) $= Q^n$ read as Q to the nth power.

Note that $Q^1 = Q$. We can also see that

$$Q^m \times Q^n = Q^{m+n}$$

This is certainly true when m and n are integers (i.e. whole numbers) greater than zero. It is also true for *all* integers, positive, negative and zero, provided that we define

$$Q^0 = 1$$

and we see then that Q to a negative power must be given by

$$Q^{-n} = \frac{1}{Q^n}$$

so that

$$1 \times Q^m \times Q^{-n} = Q^{0+m-n} = Q^{m-n} = Q^m/Q^n$$

Suppose now that m = n. We have

$$Q^m \times Q^m = Q^{m+m} = Q^{2m}$$

We also could write this as

$$Q^m \times Q^m = (Q^m)^2 = Q^{2m}$$

More generally, we can write

$$(Q^m)^n = Q^{mn}$$

Now provided that Q itself is greater than zero, *all* of these operations also work even when m and n are *not* integers, although of course it is not then possible to interpret Q^p as a product of Q taken p times. We can write for example that

$$Q^{0.5} \times Q^{0.5} = Q^{0.5 + 0.5} = Q^1 = (Q^{0.5})^2$$

and in this case it is easy to see that $Q^{1/2} = Q^{0.5}$ is the square root of Q. The interpretation is not so clear when we write that

$$Q^{0.302} \times Q^{0.477} = Q^{0.302 + 0.477} = Q^{0.779}$$

but the manipulations are just as easy.

In this connection, it is particularly interesting and useful to consider the case Q = 10, which is the *base* of our number system, and which allows us to define "common logarithms". If we write that a number A is given by

$$A = 10^a$$

then the exponent a is called the logarithm of A:

a = log A sometimes written $\log_{10} A$ to make explicit the role of the "base" 10. Note then that if b = log B, $B = 10^b$, the product of the two numbers A and B is simply the sum of their logarithms a and b.

$$\log (A \times B) = \log (10^a \times 10^b) = \log (10^{a+b}) = a + b$$
$$= \log A + \log B$$

This property was very helpful before the days of desk and pocket calculators, transforming multiplications and divisions into additions and subtractions. Logarithms are still useful in representing natural phenomena that are multiplicative rather than additive in nature. Thus, the principal utility of logarithms nowadays is in plotting quantities which vary enormously. For example, in Fig. 6-6, we use the logarithm of the atmospheric particle density to show how these densities vary with altitude, and this allows us to get significant information about both the lower and the upper atmosphere onto the same graph. On an ordinary "linear" scale, all the upper-level densities would appear squeezed down to zero by comparison with the lower-level density, which is larger by several powers of ten.

Although we make rather little use of logarithms as such in this book, we shall use whole-number exponents of ten very often indeed. We note that $10^0 = 1$, $10^1 = 10$, $10^2 = 100$, and that $10^m = 1$ followed by m zeros. Also, $10^{-1} = 1/10 = 0.1$, $10^{-2} = 1/10^2 = 1/100 = 0.01$, $10^{-3} = 0.001$, and $10^{-n} = 1/10^n$ and is equal to the number 1, n places after the decimal point, preceded by (n − 1) zeros. Indeed when we write a number such as 273.15 we see that the position of each digit with respect to the decimal point corresponds to a power of ten, and so

$$273.15 = 2 \times 10^2 + 7 \times 10^1 + 3 \times 10^0 + 1 \times 10^{-1} + 5 \times 10^{-2}.$$

Of course it is much more convenient to write 273.15. However, when we deal with numbers that are either very very large or very very small, as is often the case in astrophysics, the powers-of-ten notation allows us to write them in a particularly compact way. For example, the number of hydrogen atoms in a kilogram of hydrogen is roughly $N = 6.0 \times 10^{26}$, i.e. 6 followed by 26 zeros. The mass of an individual hydrogen atom is then

$$m_H = \frac{1}{N} = \frac{1}{6.0 \times 10^{26} \text{ atoms kg}^{-1}} = 0.166 \times 10^{-26} \text{ kg atom}^{-1}$$
$$= 1.66 \times 10^{-27} \text{ kg/atom}$$

Once one is used to this notation, it is certainly easier to handle than the "usual" notation in which m_H is written as

0.000 000 000 000 000 000 000 000 001 66 kg/atom, and N as
N = 600 000 000 000 000 000 000 000 000 atom/kg.

Also, the powers-of-ten notation allows us to express more clearly the accuracy with which we know or wish to treat a numerical value. Thus if we write $N = 6.0 \times 10^{26}$ we imply that the number is closer to 6.0 than it is to either 6.1 or 5.9 times 10^{26}, but that we do not know or care whether it is closer to 6.02 or to 6.03 times 10^{26}. With the number written out in full length as above, one might be tempted to think that all the zeros are meaningful, which is certainly not so; they simply are there to indicate the position of the decimal point. It is legitimate to write the speed of light as 299 792 458 meters per second, because all of the figures are "significant" — the accuracy is really there. However, in approximate calculations, we use $c = 3.0 \times 10^8$ m s^{-1} for the speed of light, and the powers-of-ten notation allows us to indicate quite unambiguously the level of accuracy we are interested in.

Physical terminology, dimensions and units

As we have already seen in the case of the hydrogen atom, mass has to be expressed in a specific unit; we used the kilogram (abbreviated kg). The mass could be expressed in other units, but then its numerical value would be different. If we use different units, we can transform from one to the other by use of *conversion factors*. For example, the equations defining the conversion from kilograms to grams to pounds are

$$1 \text{ kg} = 10^3 \text{ g} = 2.205 \text{ lb}$$

If we divide these equations by the quantity 2.205 lb, we obtain

$$0.454 \text{ kg lb}^{-1} = 454 \text{ g lb}^{-1} = 1$$

and we see that a conversion factor can be thought of as a special way of writing the number 1.

Mass is only one of several types of physical properties. Other properties involve *spatial* dimensions (lengths, areas, volumes), and the *temporal* dimension (time, duration). Since areas (length^2) and volumes (length^3) can be expressed in terms of lengths, we can try to limit ourselves to the fundamental *physical* "dimensions" — length, mass, and time. Not enjoying conversions of the sort 12 inches = 1 foot, 5280 feet = 1 mile, 16 ounces = 1 pound, etc. etc., I shall make no use of the "English" system of units in this book. This anachronistic system of units has been or is being abandoned by all countries except the United States. Even in the United States there is a commitment in principle to "go metric" some day, and it is ridiculous that it is taking so long. At any rate, in this book, I work for the most part with the SI ("Système international") version of the metric system, even though I, like most astrophysicists of my generation, was trained in the "cgs" (centimeter-gram-second) version. The SI is now the internationally accepted standard.

The fundamental SI units are:

Length: 1 meter (= 100 centimeters = 39.37 inch) (1 m = 100 cm)
Mass: 1 kilogram (= 1000 grams = 2.205 pound) (1 kg = 1000 g)
Time: 1 second (1 s)

The basis of the metric system, as originally set up following the French Revolution, was to define the meter in terms of the circumference of the Earth ($C = 2\pi R = 4 \times 10^7$ m), and to define the gram as the mass of a cubic centimeter of water under standard conditions. Other units, such as the kilometer, centimeter, kilogram, milligram, etc. simply involved multiplications or divisions by powers of ten. Despite this rationalization of the units of length and mass, no rationalization was made of the traditional units of time, with 60 minutes = 1 hour, 60 seconds = 1 minute, so that the period of 24 hours (24 hours = 24 hr $\times$ 60 min/hr $\times$ 60 s/min = 86 400 seconds) was taken to be the mean solar day, i.e. the mean period of rotation of the Earth relative to the Sun (see Chapter 8). The factors of 60 are a relic from the ancient Babylonian number system. Thus originally the units of length and time in the metric system were based on the size and the rate of rotation of the Earth. Today, both the meter and the second (and of course derived units such as centimeters, kilometers, milliseconds, etc.) have been *redefined* in terms of the wavelength and the frequency (see Chapter 2) of specific spectral lines of two different atoms, and the kilogram is formally defined as the mass of a prototype kilogram kept in the International Bureau of Weights and Measures in Sèvres, a suburb of Paris. We recognize that the Earth's rate of rotation varies slightly, and that the measurements of its circumference made when the meter was first defined, were not perfectly accurate. Still, the original definitions remain fairly good approximations.

With these fundamental units, we can measure many quantities besides length, area, volume, mass and duration. Let us consider *velocity*, which is the *rate* of change of *position* (say x) with respect to *time* t. If we use the Greek letter Δ (delta) to denote change, then we can write the velocity v as $v = \Delta x/\Delta t$ and we see that if position is measured in meters and time in seconds, velocity is measured in meters per second ($m\ s^{-1}$). As already mentioned, the speed of light c is roughly $c = 3.0 \times 10^8\ m\ s^{-1}$. Note that while the lower-case letter c (celerity) is *often* used as a symbol for the speed of light, this is not *always* what it stands for; the context must be checked. Obviously other units (for example miles per hour) can be used if desired, but the "dimensions" will always be length $\times$ time^{-1}. Here we have treated velocity units as derived. On the other hand, sometimes the speed of light is taken as fundamental. Thus the length unit called the *light-year* is the distance travelled at the speed of light in a year. To find its value in SI units, we simply multiply the speed of light in meters per second by the duration of a year in seconds. Thus

$$\begin{aligned} 1 \text{ light-year} &= 3.0 \times 10^8\ m\ s^{-1} \times 1\ yr \times 365\ days\ yr^{-1} \times 8.64 \times 10^4\ s\ day^{-1} \\ &= 3.0 \times 10^8\ m\ s^{-1} \times 3.2 \times 10^7\ s = 9.6 \times 10^{15}\ m \end{aligned}$$

where we have assumed that 1 year = 365 days.

Acceleration is the rate of change of *velocity* with respect to time, and it also can be expressed in terms of length and time units. Thus the acceleration a is given by

$$a = \frac{\Delta v\ (m\ s^{-1})}{\Delta t\ (s)}$$

and so the acceleration is measured in meters per second squared ($m\ s^{-2}$). For example, the acceleration of falling bodies at the Earth's surface is roughly $g = 9.8\ m\ s^{-2}$, where g is now a symbol for the gravitational acceleration, and has nothing to do with the abbreviation for gram which is written exactly the same way.

That velocity and acceleration can be expressed in terms of length and time units should hardly be surprising. It may be less obvious how we can express the units of *force*. To do this we note that Newton's Second Law of Motion states that

Force = Mass × Acceleration.

Thus, in SI units, force is in fact measured in "newtons", but the newton can be expressed in terms of meters, kilograms and seconds, with

$$1\ \text{newton} = 1\ kg \times 1\ m\ s^{-2} = 1\ kg\ m\ s^{-2}$$

To a physicist, *work* corresponds to a *force* acting on an object *and displacing it*, with

Work = Force × Displacement

Energy is the capacity to do work, and is measured in the same units. The SI unit of energy or work is called the "joule", with

$$\begin{aligned} 1\ \text{joule} &= 1\ \text{newton} \times 1\ m \\ &= 1\ kg\ m\ s^{-2} \times 1\ m = 1\ kg\ m^2\ s^{-2}. \end{aligned}$$

Note that Einstein's famous equation $E = m\ c^2$ respects these "dimensions" as of course it must. If we were to convert 1 kilogram of matter *completely* into energy, we would obtain

$$\begin{aligned} E &= 1\ kg \times (3 \times 10^8\ m\ s^{-1})^2 \\ &= 1\ kg \times 9 \times 10^{16}\ m^2\ s^{-2} = 9 \times 10^{16}\ kg\ m^2\ s^{-2} \\ &= 9 \times 10^{16}\ \text{Joules} \end{aligned}$$

I shall not prove it here, but this energy would be sufficient to lift a mass of 10^{11} kg (or 10^8 tons) to an altitude of nearly 100 kilometers above the surface of the Earth.

Energy can exist in many forms, for example chemical, nuclear, electrical, kinetic (motion) or thermal (heat). When we use energy in our day-to-day lives, we are generally changing it from one form to another. The *rate* with respect to time at which these energy transformations are being carried out is called *power* by the physicist, and the SI unit of power is in fact quite familiar, since it is the *watt*.

$$\begin{aligned} 1\ \text{watt} &= 1\ \text{joule per second} = 1\ kg\ m^2\ s^{-2}\ s^{-1} \\ &= 1\ kg\ m^2\ s^{-3}. \end{aligned}$$

Noting that a kilowatt is 10^3 watts, we see that our familiar *energy* unit, the kilowatt-hour, is given by

$$1 \text{ kWh} = 10^3 \text{ W} \times 3.6 \times 10^3 \text{ s} = 3.6 \times 10^6 \text{ joules}$$
$$= 3.6 \times 10^6 \text{ kg m}^2 \text{ s}^{-2}.$$

Indeed, we could also call a joule a watt-second. The energy of 9×10^{16} joules which we found earlier as corresponding to the mass of 1 kilogram, comes out to be 2.5×10^{10} kilowatt-hours. The average annual *per capita* energy consumption in the United States is roughly 10^5 kilowatt-hours, so that this energy corresponds roughly to the average annual energy consumption of an American city of some 250 000 inhabitants. However, even in the most advanced nuclear reactors, far more than 1 kg of fuel is involved in supplying such energy, since only a very small fraction of the mass is actually converted into energy in the case of nuclear fission.

Derived units and prefixes

We have already encountered the prefix "kilo", corresponding to a multiplying factor of one thousand (10^3), as in *kilometer* or *kilogram*. (In Europe, the word *kilo* is often used by itself, in place of kilogram.) This prefix is only one of a family of prefixes standing for multiplication by specific powers of ten. To each such prefix corresponds a symbol for use in abbreviations.

Prefix	Symbol	Factor	Prefix	Symbol	Factor
Tera	T	10^{12}	Pico	p	10^{-12}
Giga	G	10^9	Nano	n	10^{-9}
Mega	M	10^6	Micro	μ	10^{-6}
Kilo	k	10^3	Milli	m	10^{-3}
Hecto	h	10^2	Centi	c	10^{-2}
Deka	D	10^1	Deci	d	10^{-1}

Geographical distances are conveniently given in kilometers (km), while infrared wavelengths are in the micrometer (μm range); sometimes the micrometer is called the micron, and represented by the Greek letter μ standing alone. A typical wavelength in the visible might be given as 0.5461 μm or as 546.1 nm, although many physicists still work in terms of Ångstrom units (1 Å = 10^{-10} m) in which case this wavelength would be 5461 Å. Note that the megagram (1 Mg = 10^6 g = 10^3 kg) is more commonly called the (metric) ton. When we consider electromagnetic radiation (Chapter 2) we often use *frequency*, measured in cycles per second (cycles s^{-1}) or *Hertz* (Hz); often kilohertz (kHz) or megahertz (MHz) are more convenient, as in the radio-frequency domain.

Other fundamental units

Not all physical quantities can be given in terms of meters, kilograms and seconds. The property called *temperature* is measured in units of (degrees) Kelvin, abbreviated K, with 273.15 K = 0° Centigrade = 32° Fahrenheit, and 373.15 K = 100°C = 212°F. The zero of the Kelvin temperature scale is an *absolute* zero; if we think of temperature as a measure of the agitation of the microscopic constituents (molecules, atoms — see Chapter 2) of matter, then at absolute zero, there is *no* agitation. Actually, quantum effects make this statement not quite accurate, but we shall not discuss this further.

Electrical charge or current also require units independent of those discussed so far. The SI unit for current is the *ampere*, the corresponding unit of charge the *coulomb*, with 1 coulomb = 1 ampere-second. The volt will then be defined by the condition 1 volt $\times$ 1 ampere = 1 watt. We shall not go into further detail on this. Let us note, however, that the energy unit called the *electron-volt* (eV) is often used when phenomena at the atomic scale are discussed (again, see Chapter 2). The conversion factor is easily obtained when we note that the charge of an electron is 1.6×10^{-19} coulomb = 1.6×10^{-19} ampere-seconds. Then

$$\begin{aligned}1 \text{ electron-volt} &= 1.6 \times 10^{-19} \text{ ampere-seconds} \times 1 \text{ volt}\\ &= 1.6 \times 10^{-19} \text{ ampere-volt-seconds}\\ &= 1.6 \times 10^{-19} \text{ watt-seconds} = 1.6 \times 10^{-19} \text{ joule.}\end{aligned}$$

In nuclear and "high-energy" physics, we are dealing with MeV and GeV *per particle*, these particles having masses between 10^{-30} and 10^{-25} kg, so that these energies, expressed in terms of joules per kilogram, are indeed high.

Quite different units may be convenient in specific contexts. Thus in astrophysics, stellar masses are often given in units of the Sun's mass, with 1 solar mass (1 $M_\odot$) equal to 3.3×10^5 Earth masses or 2×10^{30} kg. Planetary distances are given in "Astronomical Units" (the mean Earth—Sun distance, 1 A.U. = 1.5×10^{11} m), stellar distances are measured in *parsecs* (abbreviated pc; 1 pc = 206265 A.U. = 3×10^{16} m). Further examples are given in Appendix 2, where we go from the scale of Man to the scale of the Universe. Generally, the most "reasonable" choice of unit depends on the method of measurement or calculation. In some circumstances, the choice of units is totally irrelevant. This is when we are dealing with ratios of two quantities having the same physical dimension, i.e. with "pure" or "dimensionless" numbers. An example is the number represented by the Greek letter pi, π = 3.14159265 . . . which is the ratio of the circumference of a circle to its diameter (in Euclidean geometry). Since both of these quantities are lengths, the units are irrelevant — we can use meters, miles, light-years, or whatever. Other examples are mentioned in Chapter 3.

Some physical constants

My intention here is simply to illustrate the constants which have been referred to in the text, and not to give a complete and authoritative tabulation. These constants are c, h, and σ.

Speed of light $c = 3\ 10^8\ \mathrm{m\ s^{-1}}$

The frequency ν corresponding to wavelength $\lambda = 0.3\ \mu\mathrm{m}$ is (cf. Ch.2)

$$\nu = \frac{c}{\lambda} = \frac{3\ 10^8\ \mathrm{m\ s^{-1}}}{0.3\ 10^{-6}\ \mathrm{m}} = 10^{15}\ \mathrm{s^{-1}} = 10^{15}\ \mathrm{Hz}$$

Planck's constant $h = 6.62\ 10^{-34}$ Joule $\mathrm{Hz^{-1}}$
$= 4.14\ 10^{-15}\ \mathrm{eV\ Hz^{-1}}$

Then (Ch. 2) the photon energy E corresponding to wavelength λ = 0.3 μm and frequency $\nu = 10^{15}$ Hz is

$$E = h\nu = 6.62\ 10^{-34}\ \mathrm{J\ Hz^{-1}} \times 10^{15}\ \mathrm{Hz} = 6.62\ 10^{-19}\ \mathrm{J}$$
$$= 4.14\ 10^{-15}\ \mathrm{eV\ Hz^{-1}} \times 10^{15}\ \mathrm{Hz} = 4.14\ \mathrm{eV}.$$

Stefan-Boltzmann constant $\sigma = 5.67\ 10^{-8}\ \mathrm{Wm^{-2}\ K^{-4}}$

Thus, the flux F corresponding to a temperature T = 300 K is

$$F = \sigma T^4 = 5.67\ 10^{-8}\ \mathrm{Wm^{-2}\ K^{-4}} \times (3\ 10^2\ \mathrm{K})^4$$
$$= 5.67\ 10^{-8}\ \mathrm{Wm^{-2}\ K^{-4}} \times 81\ 10^8\ \mathrm{K^4} = 459\ \mathrm{Wm^{-2}}.$$

APPENDIX II. SCALES OF THE ENVIRONMENT

		Typical size		Typical mass		Density	
		meters	other units	kilograms	other units	g/cm^3	$M_\odot/pc^3$
Adult human	(height)	1.5		50		1	
Large tree	(height)	10^2		10^6		1	
Large mountain	(height)	$5\ 10^3$	5 km	10^{16}		2.8	
Planet Earth	(radius)	$6.4\ 10^6$	$R_\oplus = 6370$ km	$6\ 10^{24}$	$M_\oplus$	5.5	
Earth—Moon system	(radius)	$3.8\ 10^8$	$60\ R_\oplus$	$6\ 10^{24}$	$(1 + 1/81)\ M_\oplus$		
SUN	(radius)	$7.0\ 10^8$	$R_\odot = 109\ R_\oplus$	$2\ 10^{30}$	$M_\odot = 3.3\ 10^5\ M_\oplus$	1.4	
Sun—Earth system	(radius)	$1.5\ 10^{11}$	1 A.U.= $215\ R_\odot$	$2\ 10^{30}$	$M_\odot$		
Solar System		$1.5\ 10^{13}$	100 A.U.	$2\ 10^{30}$	$(1 + 1/750)\ M_\odot$		
STARS	(radius)	$10^{12} - 10^{16}$	$10^{-1} - 10^3\ R_\odot$	$10^{29} - 10^{32}$	$0.05—50.0\ M_\odot$	$10^{-8} - 10^2$	
Except white dwarfs	(radius)	10^7	$R_\oplus = 10^{-2}\ R_\odot$	10^{30}	$M_\odot$	10^6	
and neutron stars	(radius)	10^4	10 km = $10^{-5}\ R_\odot$	10^{30}	$M_\odot$	10^{15}	
Interstellar medium (in galactic disk)	(dist. betw. stars)	$3\ 10^{16}$	1 pc = $2\ 10^5$ AU			10^{-23}	
The Galaxy	(diameter)	10^{21}	33 kpc	$4\ 10^{41}$	$M_g = 2\ 10^{11}\ M_\odot$	10^{-24}	10^{-2}
The Local Group	(diameter)	$3\ 10^{22}$	1 Mpc	$2\ 10^{42}$	$5\ M_g$	10^{-28}	$2\ 10^{-6}$
Distance to the Virgo Cluster		$7\ 10^{23}$	23 Mpc				
Local Supercluster	(diameter)	$2.4\ 10^{24}$	80 Mpc	10^{46}	$2\ 10^4\ M_g$	10^{-30}	$1.5\ 10^{-8}$
Observable Universe	(radius)	$2\ 10^{26}$	6000 Mpc	$4\ 10^{51}$	$10^{10}\ M_g$	10^{-31}	$2\ 10^{-9}$

Note how different units become "natural" for measuring different objects: for example, stellar masses and radii are best given in solar units. The last three entries depend on the distance scale adopted for the universe, i.e. the value of the Hubble constant which relates the redshift (interpreted as a velocity of expansion) to distance: we have used H = 50 kilometers per second per Megaparsec, but some researchers prefer 75 or 100. If the density given for the observable universe is correct, the universe is "open" and its expansion will never cease. Some cosmologists are eager to find "hidden mass" which will raise the density above the critical value necessary if the universe is to be closed and gravitation ultimately dominate the universal expansion.

APPENDIX III. PROPERTIES OF THE PLANETS COMPARED

	Mercury	Venus	Earth	Moon	Mars	Jupiter	Saturn	Uranus	Neptune	Pluto	Notes
Mean distance from Sun in A.U.	0.387	0.723	1.000	1.000	1.524	5.203	9.539	19.18	30.06	39.44	
Eccentricity of orbit	0.206	0.007	0.017	0.017	0.093	0.048	0.056	0.047	0.009	0.25	1
Solar flux (W/m^2)	6240 — 14400	2566 — 2638	1315 — 1408	1315 — 1408	490 — 712	45.7 — 55.4	13.4 — 16.8	3.37 — 4.07	1.5	0.6 — 1.6	2
Albedo	0.06	0.79	0.30	0.07	0.17	0.73	0.76	0.93	0.84	0.14 ?	
Effective temperature (K)	400 — 500	222	255	273	206 — 226	88	63	33	32	40 ?	3
Surface temperature (K)	100 — 620	750	200 — 320	100 — 380	160 — 300	120	90	60	50	40 ?	4
Atmospheric composition and surface pressure	— —	CO_2 90	N_2 O_2 1	— —	CO_2 0.006	H_2 He X	H_2 He X	H_2 He CH_4	H_2 He CH_4	? ?	5
Surface gravity ($m\ s^{-2}$)	3.6	8.6	9.8	1.6	3.7	26.2	11.3	11.5	11.6	?	
Period of axial rotation	59 d	243 d retrograde	1 d	27.3 d	24h 37m	9h 55m	10h 14m	11 h retrograde	16 h	6 d	6
Oblateness	0	0	0.003	0.02	0.009	0.06	0.10	0.06	0.02	?	7
Inclination of plane of rotation to plane of orbit (obliquity)	$<28^\circ$	3°	$23^\circ 27'$	$1^\circ 32'$	$23^\circ 59'$	$3^\circ 5'$	$26^\circ 44'$	82°	$28^\circ 48'$	?	8
Planet's radius in $R_\oplus$	0.38	0.96	1.000	0.27	0.53	11.0	9.2	3.7	3.5	?	
Planet's mass in $M_\oplus$	0.055	0.815	1.000	0.012	0.108	318	95	14.6	17.2	?	
Planet's density in g/cm^3 (water = 1)	5.5	5.1	5.5	3.3	3.9	1.3	0.7	1.6	2.2	?	

Source: for the most part, after Sagan (H.3).

Notes:

1. The eccentricity of the Moon's orbit around the Earth is 0.055; however, with respect to the Sun its orbit is indistinguishable from the Earth's, at the level of accuracy of this table.
2. This is the flux outside the atmosphere on a surface perpendicular to the rays from the Sun. Note the effects of orbital eccentricity.
3. For Mercury and Mars, both perihelion and aphelion values of the effective temperature are given. For the giant planets, the contribution of internal heating, which is not negligible, has not been included.
4. For the terrestrial planets and the Moon, the full surface range from coldest night to hottest day is estimated. For the giant planets, the temperature of the visible cloud deck is given, as no surface is observable.
5. No surface pressure is given for the giant planets, since the surface is unobservable. Pressure in atmospheres.
6. For the giant planets, the exact period of rotation depends on the latitude observed and the type of features followed, but the values given are representative. Similar effects arise if one determines the Earth's rotation by following cloud systems.
7. The Moon's shape is quite seriously distorted, and its oblateness is much too large for its present rate of rotation; this also indicates that lunar material is extremely rigid.
8. The Moon's equator is inclined by $1^\circ\ 32'$ to the ecliptic (its orbital plane around the Sun, for practical purposes), and by $6^\circ\ 40'$ to the plane of its orbit around the Earth. It is because the plane of the Moon's orbit is tilted by some 5° to the ecliptic that total eclipses of the Sun occur only relatively rarely.

Bibliography

Suggestions for Further Reading

AT the level of this book, one of the best sources of information about continuing developments in the different fields covered here is the monthly magazine *Scientific American*. Important news also appears quickly in the *New Scientist*, and in *Nature* and *Science* at a somewhat higher level. These three journals are weeklies, the last two containing high-level research articles and reports in all fields, often of interdisciplinary interest. Other journals of more or less general interest in science include *American Scientist, Endeavour, Sky and Telescope, Physics Today, Technology Review* and the *Bulletin of the Atomic Scientists.*

At a research level, there are various series of volumes of review articles which can be particularly useful to the scientist interested in familiarizing him or herself with developments in a new field. These include *Vistas in Astronomy* (Pergamon), and the various *Annual Reviews* — Astronomy and Astrophysics, Earth and Planetary Sciences, Ecology and Systematics, Energy — to name those of relevance here. Among the scientific journals, *Climatic Change, Icarus,* and *Tellus* contain more articles of interdisciplinary interest than most.

My selection and classification of books and articles is necessarily arbitrary; the divisions do not correspond exactly to the chapters of this book. More difficult works are marked with an asterisk.

A. Wide-ranging Books and Articles

1. Alfvén H., 1969. *Atom, Man and the Universe.* W. H. Freeman & Co., San Francisco.
2. Cloud, P., 1978. *Cosmos, Earth, and Man.* Yale University Press, New Haven and London.
3. Hoyle, F., 1957. *The Black Cloud.* Heinemann, London.
4. Hoyle, F., 1977. *Ten Faces of the Universe.* W. H. Freeman & Co., San Francisco and Reading (UK).
5. Shklovskii, I. S. and Sagan, C., 1966. *Intelligent Life in the Universe.* Holden-Day, Inc., San Francisco.

*6. U.S. National Academy of Sciences, 1965. *The Scientific Endeavor.* Rockefeller University Press, New York.

B. History and Philosophy of Science (see also A, D)

1. Burtt, E. A., 1932. *The Metaphysical Foundations of Modern Science* (2nd ed.). 1954 reprint — Doubleday Anchor, Garden City, NY.
2. Koyré, A., 1957. *From the Closed World to the Infinite Universe.* Johns Hopkins Press, Baltimore.
3. Kuhn, T. S., 1962. *The Structure of Scientific Revolutions.* University of Chicago Press, Chicago.
4. Munitz, M. K. (Ed.), 1957. *Theories of the Universe.* The Free Press, New York.

*5. Popper, K., 1962. *Conjectures and Refutations — the Growth of Scientific Knowledge.* Basic Books, New York.

C. Physics — The Laws of Nature

1. Calder, N., 1977. *The Key to the Universe.* Viking Press, New York.

*2. Feynman, R. P., Leighton, R. B. and Sands, M., 1963. The Feynman Lectures on Physics. Addison-Wesley Publ. Co., Reading (MA) and Palo Alto.

3. Van Heel, A. C. S. and Velzel, C. H. F., 1967. *What is Light*? McGraw-Hill World University Library, New York, Toronto.
4. March, R. H., 1978. *Physics for Poets* (2nd ed.). McGraw-Hill, New York.

*5. Newton, I., 1686. *Principia.* (Transl. A. Motte, rev. F. Cajori — 1966 reprint). University of California Press, Berkeley.

6. Pauling, L. and Hayward, R., 1964. *The Architecture of Molecules.* W. H. Freeman & Co., San Francisco.
7. Sciama, D. W., 1969. *The Physical Foundations of General Relativity.* Doubleday Anchor, Garden City, NY.
8. Trefil, J. S., 1978. *Physics as a Liberal Art.* Pergamon Press, Elmsford, NY, and Oxford.

D. Cosmology, Space and Time (see also A, B, C, E)

1. Alfvén, H., 1966. *Worlds — Antiworlds. Antimatter in Cosmology.* W. H. Freeman & Co., San Francisco.
2. Bondi, H., 1960. *The Universe at Large.* Doubleday Anchor, Garden City.

*3. Davies, P. C. W., 1978. *Space and Time in the Modern Universe.* Cambridge University Press, Cambridge (UK).

4. Eddington, A. S., 1933. *The Expanding Universe.* Cambridge University Press. (1958 reprint — University of Michigan Press, Ann Arbor).

*5. Merleau-Ponty, J., 1965. *La Cosmologie du Vingtième Siècle.* Gallimard, Paris.

*6. Narlikar, J., 1977. *The Structure of the Universe.* Oxford University Press, Oxford.

*7. Sciama, D. W., 1971. *Modern Cosmology.* Cambridge University Press, Cambridge (UK).

8. Smart, J. J. C. (Ed.), 1964. *Problems of Space and Time.* Macmillan, New York.
9. Weinberg, S., 1977. *The First Three Minutes: A Modern View of the Origin of the Universe.* Basic Books, New York.
10. Toulmin, S. and Goodfield, J., 1965. *The Discovery of Time.* Harper & Row, New York.

E. Astronomy in General

*1. Avrett, E. H. (Ed.), 1976. *Frontiers of Astronomy.* Harvard University Press, Cambridge (MA).
2. Brandt, J. C. and Maran, S. P. (Eds.), 1977. *The New Astronomy and Space Science Reader.* W. H. Freeman & Co., San Francisco.
*3. Harwit, M., 1973. *Astrophysical Concepts.* John Wiley & Sons, New York.
4. Jastrow, R. and Thompson, M., 1977. *Astronomy: Fundamentals and Frontiers* (3rd ed.). John Wiley & Sons, New York.
5. Motz, L. and Duveen, F., 1972. *Essentials of Astronomy* (2nd ed.). Wadworth Publishing Co., Belmont, California.
*6. Unsöld, A., 1969. *The New Cosmos* (transl. P. Moore). Springer-Verlag, New York.

F. Star Formation and Evolution (see also A, E)

1. Gamow, G., 1945. *The Birth and Death of the Sun.* Pelican Books, London.
2. Jastrow, R., 1967. *Red Giants and White Dwarfs.* Harper & Row, New York.
*3. Kourganoff, V., 1973. *Introduction to the Physics of Stellar Interiors.* D. Reidel, Dordrecht.
4. Meadows, A. J., 1978. *Stellar Evolution* (2nd ed.). Pergamon Press, Oxford.
*5. Reddish, V. C., 1978. *Stellar Formation.* Pergamon Press, Oxford.
6. Reeves, H., 1968. *Stellar Evolution and Nucleosynthesis.* Gordon & Breach, New York and London.

G. Extraterrestrial Intelligence (see also A)

1. Berendzen, R. (Ed.), 1973. *Life Beyond Earth and the Mind of Man.* NASA SP-328. U.S. Government Printing Office, Washington.
2. Bracewell, R. N., 1974. *The Galactic Club.* W. H. Freeman, San Francisco.
3. Morrison, P., Billingham, J. and Wolfe, J., 1977. *The Search for Extraterrestrial Intelligence* (SETI). NASA SP-419, U.S. Government Printing Office, Washington.
4. Sullivan, W., 1964. *We Are Not Alone.* McGraw-Hill, New York.

H. The Solar System (see also E)

1. Kaufman III, W. J., 1978. *Exploration of the Solar System.* Macmillan, New York.
*2. Rasool, S. I. (Ed.), 1972. *Physics of the Solar System.* NASA SP-300, U.S. Government Printing Office, Washington.
3. Sagan, C. (Ed.), 1975. The Solar System — readings from *Scientific American.* W. H. Freeman, San Francisco.
4. Wood, J. A., 1979. *The Solar System.* Prentice-Hall, Englewood Cliffs.

I. Earth Science: General; Internal Processes

1. Bates, D. R. (Ed.), 1964. *The Planet Earth* (2nd ed.), Pergamon Press, Oxford.
2. Bloom, A. L., 1969. *The Surface of the Earth.* Prentice-Hall, Englewood Cliffs, NJ.
3. Gass, I. G., Smith, P. J. and Wilson R. C. L. (Eds.), 1972. *Understanding the Earth* (2nd ed.), Artemis Press, Horsham.
*4. Munk, W. H. and MacDonald, G. J. F., 1975. *The Rotation of the Earth* (2nd ed.), Cambridge University Press, Cambridge, UK.
5. Press, F. and Siever, R. (Eds.), 1974. Planet Earth — readings from *Scientific American.* W. H. Freeman, San Francisco.
6. Press, F. and Siever, R., 1978. *Earth* (2nd ed.). W. H. Freeman.
7. Wilson, J. T. (Ed.), 1972. Continents Adrift — readings from *Scientific American.* W. H. Freeman, San Francisco.
8. Wyllie, P. J., 1976. *The Way the Earth Works.* John Wiley & Sons, New York.

J. The Sun, Solar-Terrestrial Relations, and the Earth's Upper Atmosphere and Magnetosphere (see also E)

1. Delobeau, F., 1971. *The Environment of the Earth.* D. Reidel, Dordrecht.
*2. Geophysics Research Board, 1977. *The Upper Atmosphere and Magnetosphere.* National Academy of Sciences, Washington.
*3. Gibson, E. G., 1973. *The Quiet Sun.* NASA SP-303, U.S. Government Printing Office, Washington.
*4. Mitra, A. P., 1974. *Ionospheric Effects of Solar Flares.* D. Reidel, Dordrecht.
*5. Papagiannis, M. D., 1972. *Space Physics and Space Astronomy.* Gordon & Breach, New York and London.
6. Ratcliffe, J. A., 1970. *Sun, Earth and Radio.* McGraw-Hill World University Library, New York, Toronto.
*7. White, O. R. (Ed.), 1977. *The Solar Output and its Variation.* Colorado Associated University Press, Boulder.

K. Radiation and the Atmosphere

*1. Goody, R. M., 1964. *Atmospheric Radiation.* Clarendon Press, Oxford.
*2. Houghton, J. T., 1977. *The Physics of Atmospheres.* Cambridge University Press, Cambridge, UK.
*3. Kourganoff, V., 1952. *Basic Methods in Transfer Problems.* Clarendon Press, Oxford.
4. Minnaert, M. G. J., 1954. *The Nature of Light and Colour in the Open Air.* Dover Publ., New York.
*5. Sobolev, V. V., 1975. *Light Scattering in Planetary Atmospheres.* Pergamon Press, Oxford.

L. Atmospheric Circulation, Climate, and Weather (see also I)

1. Flohn, H., 1969. *Climate and Weather.* McGraw-Hill World University Library, New York and Toronto.
2. Goody, R. M. and Walker, J. C. G., 1972. *Atmospheres.* Prentice-Hall, Englewood Cliffs, NJ.
3. Hare, F. K., 1961. *The Restless Atmosphere* (3rd ed.). Harper & Row, New York.
*4. Lorentz, E. N., 1967. *The Nature and Theory of the General Circulation of the Atmosphere.* World Meteorological Organization, Geneva.
5. Manley, G., 1952. *Climate and the British Scene.* Wm. Collins Sons, Glasgow.
6. Oliver, J. E., 1973. *Climate and Man's Environment.* John Wiley & Sons, New York.
*7. Rasool, S. I. (Ed.), 1973. *Chemistry of the Lower Atmosphere.* Plenum Press, New York.
8. Rumney, G. R., 1970. *The Geosystem: Dynamic Integration of Land, Sea and Air.* Wm. C. Brown, Publishers, Dubuque, Iowa.
*9. Sellers, W. D., 1965. *Physical Climatology.* University of Chicago Press, Chicago.
*10. Walker, J. C. G., 1977. *The Evolution of the Atmosphere.* Macmillan, New York.

M. The Oceans (see also I)

*1. Pickard, G. L., 1975. *Descriptive Physical Oceanography* (2nd ed.). Pergamon Press, Oxford.
2. Revelle, R. (Ed.), 1969. *The Ocean — a Scientific American* book. W. H. Freeman, San Francisco.
3. Turekian, K. K., 1968. *Oceans.* Prentice-Hall, Englewood Cliffs, NJ.

N. Biology and Evolution (see also A)

*1. Darwin, C., 1859. *The Origin of Species.* Mentor Edition (1958), New American Library, New York.

2. Hutchinson, G. E. *et al.*, 1970. *The Biosphere — a Scientific American* book. W. H. Freeman, San Francisco.
3. Eiseley, L., 1957. *The Immense Journey*. Random House, New York.
4. Mayr, E. (Ed.), 1978. Evolution — an issue on the history of life. *Scientific American 239*, No. 3.
5. Oparin, A. I., 1938, *The Origin of Life* (transl. S. Morgulis), Dover Publ. Co., (1953), New York.

O. Climatic Change (see also L, P, Q)

1. Brooks, C. E. P., 1949. *Climate through the Ages* (2nd ed.). Dover (1970 reprint), New York.
2. Goudie, A. S., 1977. *Environmental Change.* Clarendon Press, Oxford.
3. Gribbin, J., 1976. *Forecasts, Famines and Freezes.* Walker & Co., New York.
*4. Gribbin, J., (Ed.), 1978. *Climatic Change.* Cambridge University Press, Cambridge, UK.
*5. Lamb, H. H., 1972. *Climate: Present, Past and Future.* Methuen, London.
6. Le Roy Ladurie, E., 1971. *Times of Feast, Times of Famine.* Doubleday, Garden City, NY.
7. Schneider, S. H. and Mesirow, L. E., 1976. *The Genesis Strategy: Climate and Global Survival.* Plenum Press, New York.
*8. U.S. Comm. for GARP, 1975. *Understanding Climate Change: A Program for Action.* National Academy of Sciences, Washington.

P. Man's Impact on the Environment (see also A, Q, R)

1. Ehrlich, P. R., Ehrlich, A. H. and Holdren, J. P., 1977. *Ecoscience: Population, Resources, Environment.* W. H. Freeman, San Francisco.
*2. Holdgate, M. W. and White, G. F., 1973. *Environmental Issues.* SCOPE Report 10, John Wiley & Sons, London, New York.
3. SCEP, 1970. *Man's Impact on the Global Environment* (Report on the Study of Critical Environmental Problems). MIT Press, Cambridge, MA.
4. SMIC, 1971. *Inadvertent Climate Modification* (Report of the Study of Man's Impact on Climate). MIT Press, Cambridge, MA.
*5. Williams, J. (Ed.), 1978. *Carbon Dioxide, Climate and Society.* (Proc. IIASA Workshop), Pergamon Press, Oxford.
6. Woodwell, G. M., 1978. The Carbon Dioxide Question. *Scientific American 238*, No. 1, 34—43.

Q. Energy (see also A, P, R)

*1. Bolin, B., 1977. The Impact of Production and Use of Energy on the Global Climate. *Annual Rev. of Energy, 2,* 197—226.
*2. Geophysics Study Board, 1976. *Energy and Climate.* National Academy of Sciences, Washington.

3. Metz, W. D. and Hammond, A. L., 1978. *Solar Energy in America.* American Association for the Advancement of Science, Washington.
4. *Scientific American* book, 1972. *Energy and Power.* W. H. Freeman, San Francisco.
5. Thirring, H., 1958. *Energy for Man* (2nd ed.), Harper & Row, (reprint 1976), New York.

R. Problems of the Future of Man (see also A, O, P, Q)

1. Cipolla, C. M., 1974. *The Economic History of World Population.* Penguin Books, Harmondsworth.

*2. Clarke, J. I., 1971. *Population Geography and the Developing Countries.* Pergamon Press, Oxford.

3. Dubos, R., 1970. *Reason Awake: Science for Man.* Columbia University Press, New York.
4. Hardin, G., 1972. *Exploring New Ethics for Survival: The Voyage of the Spaceship Beagle.* Viking Press, New York.
5. Klatzmann, J., 1975. *Nourrir dix milliards d'hommes?* Presses Universitaires de France, Paris.
6. Meadows, D. H., Meadows, D. L., Randers, J. and Behrens, W. W., 1972. *The Limits to Growth* (report for the Club of Rome). Potomac Associates, Washington.
7. Mesarovic, M. and Pestel, E., 1974. *Mankind at the Turning Point.* (2nd report to the Club of Rome). E. P. Dutton, New York.
8. Tinbergen, J., 1976. *Pour une Terre Vivable.* Elsevier Sequoia, Bruxelles, and Editions Québec-Amérique, Ottawa.
9. Tinbergen, J. (coord.), 1976. *Reshaping the International Order.* (report to the Club of Rome). E. P. Dutton, New York.
10. Wortman, S. *et al.*, 1976. Food and Agriculture, a special issue of *Scientific American, 235,* No. 3.

Selected Notes and References

References to works listed in the *Suggestions for Further Reading* are given in the form (B.4), which refers to the anthology by Munitz. When possible, I refer to books or articles at the level of this book (for example, a *Scientific American* article), rather than to the original research publication. A complete bibliography would have been prohibitively long; further references can be found in the articles and books listed here and above.

Chapter 1

1. Tournier, Michel, 1975. *Les Météores.* Gallimard, Paris.
2. Cf. the volume containing both CNRS Colloque Internat. No. 263 — L'Evolution des Galaxies et ses Implications Cosmologiques, and

IAU Coll. 37, Décalages vers le Rouge et Expansion de l'Univers, Editions du CNRS, Paris, 1977, especially the discussions by G. Burbidge, M. J. Rees and P. Morrison, pp. 555—612. Also J. C. Pecker, Possible explanations of non-cosmological redshifts, in the above, and H. Alfvén. La Cosmologie, Mythe ou Science? in La Recherche No. 69, 610—616, 1976.

Chapter 2

1. Halley, E., 1686. Ode dedicated to Newton. Transl. L. J. Richardson, in (C.5), vol. 1, p. xiii.
2. Hawking, S. W., 1977. The Quantum Mechanics of Black Holes. *Sci. Amer. 236*, No. 2, 34—40.
3. Marshak, R. E., 1969. The Fourth Force in Nature. *Amer. Scientist 57*, 517—535.
4. Freedman, D. Z., van Nieuwenhuizen, P., 1978. Supergravity and the Unification of the Laws of Physics. *Sci. Amer. 238*, No. 2, 126—143.

Chapter 3

1. Mach, E., in (D.8), pp. 126—131.
2. Alpher, R. A., 1973. Large Numbers, Cosmology, and Gamow. *Amer. Scientist 61*, 52—58. Also E. R. Harrison, 1972. The Cosmic Numbers. *Phys. Today 27*, No. 2, 30—37.
3. Van Flandern, T. C., 1976. Is Gravity Getting Weaker? *Sci. Amer. 234*, No. 2, 44—52. Cf. also P. J. Smith, 1978. The end of the expanding Earth hypothesis? *Nature 271*, 301.
4. Layzer, D., 1975. The Arrow of Time. *Sci. Amer. 233*, No. 6, 59—69.
5. Harrison, E. R., 1972. Why the Sky is Dark at Night. *Phys. Today, 27*, No. 2, 30—37.
6. Muller, R. A., 1978. The Cosmic Background Radiation and the New Aether Drift. *Sci. Amer. 238*, No. 5, 64—74.
7. Newton, in (C.5), vol. 1, p. 6.
8. Koyré, in (B.2), ch. X.
9. Sciama, (C.7), ch. 3.

Chapter 4

1. Huygens, C., 1698. *Cosmotheoros.* in (B.4), p. 222.
2. Cameron, A. G. W., 1973. Abundances of the Elements in the Solar System. *Space Sci. Rev. 15*, 121—146.
3. Langer, G. E., Kraft, R. P. and Anderson, K. S., 1974. FG Sagittae — The s-process episode. *Astrophys. J. 189*, 509—522. Cf. Also I. J. Christy-Sackmann, K. H. Despain, 1974. An Interpretation of the Puzzling Observations of FG Sagittae. *Astrophys. J. 189*, 523—530.
4. Schramm, D. N. and Clayton, R. N., 1978. Did a Supernova Trigger the Formation of the Solar System? *Sci. Amer. 239*, No. 4, 98—113.

5. He was condemned by the Inquisition in Rome and burned at the stake in 1600.

Chapter 5

1. de Saint-Exupéry, A., 1943. *Le Petit Prince.* Harcourt, Brace & World, New York, Gallimard, Paris.
2. Shapley, D., 1977. Ocean scientists may wash hands of Sea Law Treaty. *Science 197,* 645. Cf. also J. A. Knauss, 1974. Marine Science and the 1974 Law of the Sea Conference. *Science 184,* 1335—41.
3. Eddy, J., 1976. The Maunder Minimum. *Science 192,* 1189—1202. also, 1977. Climate and the Changing Sun. *Climatic Change 1,* 173—190. 1977. The Case of the Missing Sunspots. *Sci. Amer. 236,* No. 5, 80—92.
4. CIA, 1974. Potential implications of trends in world population, food production and climate. Report OPR-401, Central Intelligence Agency, available through DOCEX project, Library of Congress, Washington. Cf. also (O.7) and (R.10).
5. Mitchell, Jr., J. M., in (Q.2), pp. 53, 55. Also M. I. Budyko, The effect of solar radiation variations on the climate of the Earth. *Tellus 21,* 611—619. and GARP Publ. No. 16 — The Physical Basis of Climate and Climate Modeling. World Meteorological Organization, Geneva. Also W. S. Dansgaard *et al.*, 1971. Climatic record revealed by the Camp Century Ice Core, in K. Turekian (Ed.), The Late Cenozoic Glacial Ages, Yale Univ. Press, New Haven, pp. 37—56.
6. See (I.3), pp. 89—95. Also D. Pines and J. Shaham, 1973. Seismic Activity, Polar Tides, and the Chandler Wobble, *Nature 245,* 77—81.
7. Lorius, C. and Duplessy, J. C., 1977. Les grands changements climatiques *La Recherche 8,* 947—955. Cf. also N. J. Shackleton, N. D. Opdyke, 1976. Geol. Soc. Amer. Memoir., 145 and N. J. Shackleton *et al.* (Eds.) Initial reports of the Deep Sea Drilling Project.
8. Rasool, I. and De Bergh, C., 1970. The runaway greenhouse and the accumulation of CO_2 in the Venus atmosphere. *Nature 226,* 1037—1039.

Chapter 6

1. Copernicus, N., 1543. De Revolutionibus Orbium Coelestium, 1.1, cap. X. (transl. A. Koyré). (B.2), p. 33. Cf. also (B.4), pp. 149—173.
2. Bahcall, J. N. and Sears, R. L., 1972. Solar neutrinos. *Ann. Rev. Astron. Astrophys. 10,* 25—44.
3. Opik, E. J., 1958. Climate and the Changing Sun. *Sci. Amer. 198,* 85—92. Dilke, F. W., Gough, D. O., 1972. The Solar Spoon. *Nature 240,* 262—3.
4. See note 3 for Chapter 5.
5. SMIC (P.4), p. 245. Cf. also Manabe, S., in S. F. Singer (Ed.), *Global Effects of Environmental Pollution.* D. Reidel, Dordrecht, 1970. pp. 156—157.

6. King, J. W., 1973. Solar radiation changes and the weather. *Nature 245,* 443—446. and 1974 Weather and the Earth's magnetic field. *Nature 247,* 131—134.
W. O. Roberts and R. H. Olson, 1973. New evidence for effects of variable solar corpuscular emission on the weather. *Rev. Geophys. Space Phys. 11,* 731—740 and 1975. Great Plains Weather. *Nature 254,* 380. C. O. Hines and I. Halevy, 1975. Reality and nature of a Sun-weather correlation. *Nature 258,* 313—314. Dickinson, R. E., 1975. Solar variability and the lower atmosphere. *Bull. Amer. Meteorol. Soc. 56,* 1240—1248. For a critical summary, see Pittock, A. B., 1978. A Critical Look at Long-Term Sun-Weather Relationships. *Rev. Geophys. Space Phys. 16,* 400—440.

Chapter 7

1. Lucretius, ca. B.C. 60. cf. (B.4) pp. 41—97, or *The Nature of the Universe,* transl. R. E. Latham, Penguin Books, London, 1951.
2. Preliminary results appeared in *Science 203,* 23 February 1979.
3. Ellis, J. S. and Vonder Haar, T. H., 1976. Zonal Average Earth Radiation Budget Measurements from Satellites for Climate Studies. Colorado State Univ. Atmosph. Sci. Paper No. 240, Fort Collins.

Chapter 8

1. Comte, A., 1835. Cours de Philosophie Positive. T. II — La Philosophie Astronomique et la Philosophie de la Physique. Bachelier, Paris, pp. 32—33.
2. Berger, A., 1977. Long-term variation of the Earth's Orbital Elements. *Cel. Mech. 15,* 53—74. And 1978. Long-term variations of caloric insolation resulting from the Earth's orbital elements. *Quatern Res. 9,* 139—167.
3. Milankovitch, M., 1920. Théorie Mathématique des Phénomènes Thermiques produits par la Radiation Solaire. Gallimard, Paris. Mason, B. J., 1976. Towards the understanding and prediction of climatic variations. *Quart. J. R. Meteor. Soc. 102,* 473—498. Hays, J. D., Imbrie, J., Shackleton, N. J., 1976. Variations in the Earth's Orbit — Pacemaker of the Ice Ages. *Science 194,* 1121—1132. Berger, A., 1977. Support for the astronomical theory of climatic change. *Nature 269,* 44—45.
4. Ward, W. R., 1973. Large-scale variations in the obliquity of Mars. *Science 181,* 260—262.
5. Gribbin, J. and Plagemann, S., 1974. *The Jupiter Effect.* Walker & Co; New York.
6. Kant, I., 1754. Examination of the question whether the Earth has undergone an alteration of its axial rotation. In Universal Natural History and Theory of the Heavens (transl. W. Hastie), Univ. of Michigan Press, Ann Arbor, 1969.

7. See (I.3), (I.4), and Kahn, P. G. and Pompea, S. M., 1978. — Nautiloid growth rhythms and dynamical evolution of the Earth—Moon system. *Nature 275*, 606—614.

Chapter 9

1. Shelley, P. B., 1820. 'Ode to the West Wind'.
2. Young, R. E. and Pollack, J. B., 1977. A Three-dimensional Model of Dynamical Processes in the Venus Atmosphere. *J. Atmos. Sci. 34*, 1315—1351.
3. Ingersoll, A. P., 1976. The Meteorology of Jupiter. *Sci. Amer. 234*, No. 3, 46—56.

Chapter 10

1. Aeschylus, ca. B.C. 470. *Prometheus Bound.* Transl. P. E. More.
2. Bretherton, F. P., 1975. Recent developments in dynamical oceanography. *Quart. J. R. Meteorol. Soc. 101*, 705—721.
3. Schneider, S. H. and Dickinson, R. E., 1974. Climate Modelling. *Rev. Geophys. Space Phys. 12*, 447—493. Cf. also GARP Publ. No. 16.

Chapter 11

1. Dumas, J. and Boussingault, J. B., 1844. *Essai de Statistique Chimique des Etres Organisés*, 3 ème éd., Fortin, Masson et Cie., Paris, p. 11. also Oparin (N.5), p. 245.
2. Margulis, L. and Lovelock, J. E., 1974. Biological Modulation of the Earth's Atmosphere. *Icarus 21*, 471—489.
3. Cf. Note 8, Chapter 5.
4. WMO, 1977. Report of the Scientific Workshop on Atmospheric Carbon Dioxide. WMO No. 474, World Meteorological Organization, Geneva.
5. Cloud, P. and Gibor, A., 1970 in (N.2), p. 63.
6. See (P.2); also Delwiche, C. C., in (N.2), 69—80.

Chapter 12

1. Thoreau, H. D., 1854. *Walden, or Life in the Woods.* Houghton Mifflin edition, 1897, Boston and New York, pp. 312—313.
2. Otterman, J., 1977. Anthropogenic Impact on the Albedo of the Earth. *Climatic Change 1*, 137—155. Also Landsberg, H. E., 1970. Man-made climatic changes. *Science 170*, 1265—1274.
3. See (P.6); also Woodwell, G. M. *et al.*, 1978. The Biota and the World Carbon Budget. *Science 199*, 141—146. And Bolin, B., 1977. Changes of Land Biota and their importance for the Carbon Cycle. *Science 196*, 613—15. Also (P.5).

4. Cf. e.g. Bryson, R. A., 1974. A Perspective on Climate Change. *Science 184*, 753—760.
5. Keeling, C. D. *et al.*, 1976. Atmospheric Carbon Dioxide Variations at Mauna Loa Observatory, Hawaii. *Tellus 28*, 538—551.
6. Bach, W., 1976. Global Air Pollution and Climatic Change. *Rev. Geophys. Space Phys. 14*, 429—474.
7. Schipper, L., 1976. Raising the Productivity of Energy Utilization. *Ann. Rev. Energy 1*, 455—517.
8. Cf. Landsberg, Note 2 above; also Dettwiller, J., 1978. L'évolution séculaire de la température à Paris. La Météorologie, VI° sér., No. 13, 95—130.
9. Cf. SMIC (P.4).
10. Fabian, P., 1978. Ozone increase from Concorde operations. *Nature 272*, 306—307. Cf. also Thrush, B. A., 1978. Recent developments in atmospheric chemistry. — *Nature 276*, 345—347. These findings should be compared with the 1975 picture, as given e.g. in Machta, L., 1976. The Ozone Depletion Problem (An Example of Harm commitment), MARC Report No. 1, Chelsea College, London.

Chapter 13

1. Frost, R., 1923. 'Fire and Ice', from *The Poetry of Robert Frost* edited by Edward Connery Latham. Copyright 1923, © 1979 by Holt, Rinehart & Winston. Copyright 1951 by Robert Frost. Reprinted by permission of Holt, Rinehart and Winston, Publishers, and Jonathan Cape Limited.
2. Livingstone, D. A., 1971. Speculations on the climatic history of mankind. *Amer. Scientist 59*, 332—337. Also, Ardrey, R., 1976 — The Hunting Hypothesis. Atheneum, New York, tells an interesting story.
3. Based on (Q.2), Table 1.6, p. 43, and (R.1), Table 14b.
4. Brown, H., 1976. Energy in our Future. *Ann. Rev. Energy 1*, 1—36. also Thirring, 1956 (Q.5), ch. X, and of course many other publications of recent years.
5. Mercer, J. H., 1978. West Antarctic ice sheet and CO_2 greenhouse effect; a threat of disaster. *Nature 271*, 321—325.
6. Durrani, S. A., 1978. Natural fission reactors — Oklo style. *Nature 271*, 306—307. Cowan, G. A., 1976. A Natural Fission Reactor. *Sci. Amer. 235*, No. 1, 36—47.
7. Medvedev, Zh., 1977. Facts behind the Soviet nuclear disaster. *New Scientist 74*, 761—764.
8. Maimonides, M., (transl. M. Friedlander). *The Guide for the Perplexed.* Dover Publ. Co., New York, 1956. p. 275.
9. Glaser P., 1977. Solar Power from Satellites. *Phys. Today 30*, No. 2, 30—38. Cf. also correspondence in *30*, No. 7.
10. Mendillo, M., Hawkins, G. S. and Klobuchar, J. A., 1975. A large-scale hole in the ionosphere caused by the launch of Skylab. *Science 187*, 343—345. Also Lewis, R. S., 1978. Environmental impacts of the space shuttle. *New Scientist*, 16-2-78, 414.
11. O'Neill, G. K., 1974. The Colonization of Space. *Phys. Today 27*, No. 9, 32—40.

Index of Names

Index of Subjects